D1040911

# NOTES

# NOTES

# NOTES

# FYI - Information about NOAA Weather Radio

**NOAA Weather Radio** broadcasts National Weather Service (NWS) warnings, watches, forecasts, and other non-weather related hazard information 24 hours a day. During an emergency, NWS sends a special tone that activates weather radios in the listening area. Weather radios equipped with a special alarm tone feature can sound an alert and give you immediate information about a life-threatening situation.

NOAA Weather Radios are found in many electronics stores and cost about $25-$100. Some features to consider are alarm tone, battery backup, and "Specific Area Message Encoding" (SAME) programming.

NOAA Weather Radio (NWR) broadcasts warnings and post-event information for all types of hazards - **weather** *(blizzards, winter/ice storms and thunderstorms)*, **natural** *(floods, hurricanes, tornadoes and earthquakes)*, **technological** *(chemical or oil spills, nuclear power plant emergencies, etc.)*, and **national emergencies**. NOAA collaborates with other Federal agencies and the FCC's Emergency Alert System (EAS) to issue non-weather related emergency messages. Weather Radio also cooperates in providing notices under the Department of Homeland Security Threat Advisory System and is also used for issuance of "Amber Alert" (kidnapped/endangered child notices.)

Local NWS offices also broadcast non-weather related emergency messages on NWR *at the request of local or state government officials*. During this type of emergency, officials provide text information about the hazard and recommend protective actions for the affected areas. NWS offices have pre-arranged agreements with government to speed the process, since minutes and seconds make a difference.

## Programming Your NOAA Weather Radio

If you purchase a Weather Radio receiver with "Specific Area Message Encoding" (SAME), you should program it with the coding for your area. By doing so, you can limit the alerts which will trigger your weather radio to only those affecting your warning area. Follow the manufacturer's directions to program your receiver, using the six-digit SAME code(s) for the warning areas of interest to you.

To learn more about **NOAA Weather Radios** visit **www.nws.noaa.gov/nwr**

*(Click on "SAME Coding" to find YOUR state, county or territory codes!)*

| ST | County/City | SAME # | NWR TRANSMITTER | FREQ. | CALL | WATTS |
|---|---|---|---|---|---|---|

# IT'S

# A

# DISASTER!

## ...and what are YOU gonna do about it?

*3rd Edition*

## A Disaster Preparedness, Prevention & Basic First Aid Manual

by Bill & Janet Liebsch

Fedhealth
7739 E. Broadway Blvd. #416
Tucson, AZ 85710-3941
info@fedhealth.net

ISBN 1-930131-11-9
Published by Fedhealth
Indexing services by Michelle B. Graye, Tucson, AZ.
Printed by Dual Printing, Inc. • www.dualprinting.com
Printed on recycled paper
3rd Edition: September 2003  (Revised October 2004)

Library of Congress Catalog Card Number: 2003091466

This Manual is available through the following methods:

- for individual purchase directly from Fedhealth

- for schools and youth groups to use as a fundraising
  project and earn <u>65%</u> profits

- for companies and organizations to customize as
  giveaways for employees, customers & communities

- for qualified Resellers to offer online and in stores!

For more information please visit Fedhealth online:

http://www.fedhealth.net

(Click on *BOOKS*, the *ULTIMATE FUNDRAISER*,
*CORPORATE SALES*, or *RESELLER PROGRAM*)

Downloadable Tools for above programs now online!

# ACKNOWLEDGEMENTS

We would like to personally thank the following individuals for their support and belief in us and in this venture over the past several years:

Stephen R. (Dex) Dexter, William H. (Bill) Holt of Sammamish, WA, and Genevieve, Richard and Ann Worley.

We also would like to thank several individuals who contributed materials, expertise and time to help us during the compilation and enhancement of this manual:

Doug Abromeit (USFS National Avalanche Center), Peter Baker (American Red Cross), Jeff Brown (American Avalanche Association), Max London (OCIPEP), Rick Meitzler (Charleston County EPD), Joe & Tamara Melanson, Chris Tucker (OCIPEP), Donna Warner (Canadian Network of Toxicology Centres, Univ. of Guelph), Patrice White (DC EMA) and various personnel from the Canadian Red Cross.

And MOST important of all…

Thank you to all our family members and friends who believed in us and supported us spiritually and emotionally!

## ABOUT THE AUTHORS

Bill and Janet Liebsch are the original founders of Fedhealth, a publishing and marketing corporation formed to help the public focus on preparedness and health-related issues. They consider themselves "social entrepreneurs" dedicated to developing and marketing programs that primarily benefit schools and nonprofit organizations.

Fedhealth's manuals will be continually updated on preparedness and safety-related topics and expanded to include other languages.

## DISCLAIMER

The authors of this Manual are not licensed physicians, and the enclosed suggestions should not replace the medical advice of trained medical staff. This information is not intended as a substitute for a first aid course, but reviews some basic first aid measures that could be used when professional medical assistance is delayed or temporarily unavailable due to a major disaster or crisis.

# DEDICATION

*This manual is dedicated to Volunteers all around the
world who give their heart, soul, energy, and time
unselfishly for the betterment of our society.
Thank you.*

# CONTENTS

# INTRODUCTION

If you have never been involved in any type of major disaster, count yourself among the lucky ones and realize that disasters can happen anywhere and anytime!!

As victims of disasters can confirm, the confusion immediately following a disaster is <u>scary</u> - especially if you have not prepared yourself in advance and discussed these ideas with your family members.

Hopefully every time you see or hear about a disaster it makes you stop and think… "What if that was me or my family?"  But what have <u>YOU</u> done to get yourself and your family ready?  The best thing you can do to deal with any type of disaster is…

<u>**BE AWARE**</u>… <u>**BE PREPARED**</u>… and… <u>**HAVE A PLAN**</u>!

If you do these 3 things, the life you save could be your own… because what you <u>don't</u> know <u>CAN</u> hurt you!

The more the public is prepared for a disaster, the less strain we place on our local emergency services.  Any major disaster will temporarily swamp first responders, therefore, both the Red Cross and the Federal Emergency Management Agency recommend persons to try to be self-sufficient for at least <u>72</u> <u>HOURS</u> following a disaster.  And if you are prepared for a longer period… that's even better!

A majority of this information was compiled from various publications  provided by the Red Cross, U.S. Department of Homeland Security and FEMA, Public Safety and Emergency Preparedness Canada, the CDC and Health Canada to help assist you in preparing for various types of disasters and basic first aid. It also will offer many suggestions on personal checklists and important telephone numbers for your family members and emergency groups that can be written in the spaces provided  or attached inside this Manual.

We realize you may not experience every type of disaster in your part of the world, but if you ever travel away from home you could potentially be placed in a disaster situation so please educate yourself and your family.

Please stop your hectic lives for just a few hours and sit down with your entire family (from children to seniors) to read this Manual and discuss how each of you would handle these types of situations.

It will be quality time with your loved ones and could save your lives!

# Disaster Facts & Figures

Both natural and man-made disasters are becoming more common all around the world. El Niño and La Niña events impact billions of people since these climate extremes disrupt jet streams and regions of high and low pressure. These disruptions can potentially increase or decrease weather-related disasters such as extreme heat and cold, floods, hurricanes and thunderstorms.

According to the National Oceanic and Atmospheric Administration's Pacific Marine Environmental Laboratory Tropical Atmosphere Ocean project, El Niño happens when tropical Pacific Ocean trade winds die out and ocean temperatures become unusually warm. La Niña occurs when the trade winds blow unusually hard and sea temperatures become colder than normal. These warm and cold phases are referred to as El Niño/Southern Oscillation, or ENSO, which has a period of roughly 3-7 years. Although ENSO originates in the tropical Pacific system, it has effects on patterns of weather all over the world.[1]

El Niño (warm episodes) occur about every 4-5 years and can last up to 12 to 18 months.[2] La Niña (cold episodes) conditions recur every few years and typically last 9 to 12 months but can persist for as long as 2 years.[3] There are also periods where the system is neither warm nor cold (called neutral conditions).

Even though there is still much to learn about these systems and their impacts on the global community, ENSO forecasts can help individuals, businesses, and governments prepare for these events.

*To learn more about NOAA's Pacific Marine Environmental Laboratory Tropical Atmosphere Ocean project visit www.pmel.noaa.gov/tao/*

In addition to the climate extremes mentioned above, some key elements in the increasing numbers of worldwide disasters include:

- Global warming and cooling trends
- Larger cities are sprawling into high-risk zones
- World population is nearly 6 billion and growing causing global water consumption to increase
- Humans are damaging our natural resources (e.g. pollution, destroying rain forests, coral reefs, wetlands, etc.)

The World Meteorlogical Organization has released evidence that the 1990s were the warmest decade globally since instrumental measurements started in the 1860s with 1998 still being the warmest year on record.

# GENERAL FACTS & FIGURES ON DISASTERS

Although there are still many unknowns, global warming and cooling trends pose real risks that could potentially alter sea levels, food and water supplies and climate conditions around the world.

According to the Worldwatch Institute, 10 million people died as a result of natural catastrophes in the 20th century.

The costs of weather-related disasters in just the United States alone average $1 billion per week!

Year after year it appears the most frequent natural disasters are windstorms and floods, which combined usually account for 80%-90% of the worldwide economic losses.

Every year hundreds of millions of people worldwide are evacuated or driven from their homes due to natural disasters.

According to Munich Re Group, (a German reinsurance company that monitors worldwide natural disasters) the following summarizes major losses around the world:

| Year | Worldwide Economic Losses | # of Major Disasters | # of deaths by Major Disasters |
|------|---------------------------|----------------------|-------------------------------|
| 2003 | $65 billion (in US $) | 700 | 75,000 |
| 2002 | $55 billion (in US $) | 700 | 11,000 |
| 2001 | $36 billion (in US $) | 700 | 25,000 |
| 2000 | $30 billion (in US $) | 850 | 10,000 |
| 1999 | $100 billion (in US $) | 755 | 75,000 |
| 1998 | $92 billion (in US $) | 700 | 50,000 |

Please note the above figures do not include economic losses caused by smaller natural disasters that occur daily around the world.

Also, note the number of known deaths from major disasters in 2003 was almost seven times higher than 2002 due to a deadly earthquake in southeastern Iran. This dramatic statistic proves how devastating disasters can be in highly populated areas.

Now let's look at some facts and figures on specific types of disasters to get a better understanding of how they impact the world.

# Facts & Figures By Type Of Disaster

## Avalanches, Landslides & Mudflows

Statistics show there are about 1 million snow avalanches worldwide each year!

Flooding in Venezuela triggered landslides and mudflows that washed away entire villages and mountain slopes claiming more than 30,000 lives in 1999.

Peru experienced one of their worst landslide disasters when a 3-million-ton block of ice split from a melting glacier creating a destructive wave of ice, mud and rocks that traveled 10 miles (16 km) in just 7 minutes killing more than 4,000 people.

## Earthquakes

The U.S. Geological Survey estimates there are 500,000 detectable earthquakes in the world each year - only 100,000 are felt and 100 cause damage.

The world's deadliest earthquake on record hit central China in 1557 killing an estimated 830,000 people.

Earthquakes can happen in virtually any region in Canada although most are concentrated in the western and eastern provinces and territories.

Over 600 million people live in areas that are at risk from earthquakes.

Some of the strongest earthquakes in U.S. history (est 7.9-8.2) occurred on the New Madrid fault (general area between St. Louis and Memphis) back in 1811-1812. This area still experiences about 200 earthquakes a year.

Two of the most violent earthquakes in North America were in British Columbia's Queen Charlotte Island (8.3) and in Anchorage Alaska (9.2).

Aftershocks may be felt for several days, weeks, months or even years depending on the force of a major earthquake.

## Extreme Heat

According to NOAA's National Climatic Data Center, 80% of the contiguous United States was in moderate-to-extreme drought back in July 1934.

Today less than half a billion people live in water-stressed countries but projections indicate by 2025 that number could increase to 3 billion!

## Fire

At least 80% of all fire deaths occur in residences -- meaning homes, apartments, condos and mobile homes.

Fire kills more Americans every year than all natural disasters combined and careless smoking is the leading cause of fire deaths.

More forests burned in 1997 than at any time in recorded history. According to a report issued by the World Wide Fund for Nature, 80% of those fires were set deliberately to clear land for planting or development.

## Floods
The year 2000 floods in Mozambique left nearly 1 million people homeless and affected hundreds of thousands of people again in 2001.

More than 90% of declared disasters include flooding.

The Worldwatch Institute reports 13 of the world's 19 megacities (cities with over 10 million people) are in coastal zones -- and 2 billion (or 1 in 3) people live within 60 miles (100 km) of a coastline.

Flash floods can cause walls of water reaching heights of 20 feet (6 m).

## Hailstorms
In 1991, Calgary Alberta experienced the worst hailstorm in Canadian history when a 30-minute storm caused about $400 million in damage.

On May 22, 1986 an unusual killer hailstorm in China's Sichuan Province left 9,000 people injured and 100 dead.

The largest known hailstone ever measured in the U.S. was found in Aurora, Nebraska on June 22, 2003 with a record 7-inch (17.78 cm) diameter and a circumference of 18.75 inches (over 47 cm)!

## Hazardous Materials
As many as 500,000 products pose physical or health hazards and can be defined as "hazardous materials" and over 1,000 new synthetic chemicals are introduced each year.

Each year about 400 million metric tons of hazardous wastes are generated worldwide.

According to FEMA, varying quantities of hazardous materials are manufactured, used, or stored at an estimated 4.5 million facilities in the U.S.

## Hurricanes, Cyclones & Typhoons
Japan, China, the Philippines and other parts of Southeast and East Asia average about 20 typhoons a year.

Over 75 million Americans live in hurricane areas.

An average of 5 hurricanes strike the U.S. each year.

Nine out of 10 hurricane deaths are due to storm surge (a rise in the sea level caused by strong winds). Storm surges can get up to 20 feet (6 m) high and 50 miles (80 km) wide!

One of the worst cyclone disasters in recorded history struck Bangladesh and India killing between 500,000 and 1 million people back in 1970.

## Nuclear Power Plants / Nuclear Incident

The most immediate danger from an accident at a nuclear power plant is exposure to high levels of radiation.

Winds and weather could possibly impact people up to 200 miles (320 km) away from the accident site.

The April 1986 explosion at Ukraine's Chernobyl nuclear power plant burned for 10 days releasing about 5% of the radioactive reactor core into the atmosphere exposing millions of people to varying doses of radiation.

## Terrorism

The U.S. Department of Defense estimates that as many as 26 nations may possess chemical agents and/or weapons and an additional 12 may be seeking to develop them. *(Per FEMA's web site as of February 11, 2003)*

The Central Intelligence Agency reports that at least ten countries are believed to possess or be conducting research on biological agents for weaponization. *(Per FEMA's web site as of February 11, 2003)*

Threats or acts of terrorism cause fear and anxiety in adults and children, but don't let it consume you since it will impact your health. Learn about risks and discuss how best to handle them with the entire family. Stay current on alerts but don't obsess over the news ... and stick to your daily routine!

## Thunderstorms & Lightning

On average, the U.S. has 100,000 thunderstorms each year.

At any given moment, nearly 1,800 thunderstorms can be in progress over the face of the earth!

It is a myth that lightning never strikes the same place twice -- it often strikes the same site several times in the course of one storm.

## Tornadoes

The U.S. has more tornadoes than any other place in the world and averages 1,000 tornado sightings each year.

In 1974, during a 21-hour period, 148 tornadoes ripped through 13 states and 1 province between Alabama and Ontario, Canada killing 315 people.

Tornadoes can last for several seconds or more than an hour, but most last less than 10 minutes.

A waterspout is a tornado over water but isn't recorded until it hits land.

## Tsunamis

A tsunami [soo-nah´-mee] is a series of huge, destructive waves usually caused by an earthquake, volcanic eruption, landslide or meteorite.

A tsunami is NOT a tidal wave — it has nothing to do with the tides!

The West Coast / Alaska Tsunami Warning Center reports a 1958 landslide generated tsunami in Lituya Bay, Alaska produced a 1,722 foot (525 m) wave!

## Volcanoes

More than 65 active or potentially active volcanoes exist in the U.S. and over 40 of them are in Alaska!

According to the Catalog of Active Volcanoes published by the Smithsonian Institution there are about 850 active volcanoes that have erupted in the last few hundred years. About 600 of these volcanoes are part of the "Ring of Fire," a region that encircles the Pacific Ocean.

Volcanic eruptions can hurl hot rocks easily 20 miles (32 km) or more.

An erupting volcano can also trigger tsunamis, flash floods, earthquakes, rockfalls, landslides and mudflows.

## Winter Storms / Extreme Cold

The leading cause of death during winter storms is from automobile or other transportation accidents.

Cold weather puts an added strain on the heart. Exhaustion or heart attacks caused by overexertion (like shoveling snow or pushing a car) are the second most likely cause of winter storm-related deaths.

The risk of hypothermia is greatest among elderly persons who literally "freeze to death" in their own homes.

The Canadian ice storm of 1998 created an economic loss of almost $3 billion with massive power outages affecting over 4 million people!

# Section 1

## Family Information, Plan & Kits

# Family Information & Phone Numbers

Place these records in a safe location (such as a metal box or a safety deposit box). We suggest you review/update the information several times a year to keep records current.

Since this data changes quite often, we suggest you use the information below as a guide and write everything down on a piece of paper then paperclip inside this Manual for easy access. Keep a record of each school your child or children attend and please replace it every time there is a change. And make sure other family members get updates too!

## List work and/or school addresses & Phone numbers of all Family Members:

**Parent/Guardian works at:** _____

Work address: _____

Work & Cell Phone #s: _____

**Parent/Guardian works at:** _____

Work address: _____

Work & Cell Phone #s: _____

**Brother/Sister works at:** _____

Work address: _____

Work & Cell Phone #s: _____

## School information for each child in Family:

Child's name: _____

School name: _____

School address: _____

Main phone # for school: _____

Contact name at school: _____

- Will school HOLD or RELEASE child if an emergency or disaster?


- Where will the school move child if an emergency or disaster?

**Suggestion:** Parents and Guardians may want to keep a copy of your child or children's information at your place of employment <u>and</u> with another family member in case of a disaster or emergency. Please make sure you update records each year so everyone has the right data.

## Other Important Family Information:

Please write information down on a piece of paper and place in a safe location (such as a fireproof metal box or a safety deposit box). We also suggest you review / update information <u>several</u> times a year to keep records current.

**Make a list of <u>each</u> Family Member's Social Security Number**

Name: _____

Social Security #: _____

**HMO/Insurance Policies:**

Insurance Co. Name: _____

Policy #: _____ Phone #: _____

Insurance Co. Name: _____

Policy #: _____ Phone #: _____

Family Doctor Name _____

Family Doctor's Address _____

Dr. Phone #: _____

Closest Hospital Name _____

Closest Hospital Address _____

Hospital Phone #: _____

# FAMILY EMERGENCY PLAN CHECKLIST

The next time disaster strikes, you may not have much time to act and local first responders may not be able to reach you right away. PREPARE NOW for a sudden emergency and discuss these ideas with your entire family to create a **Family Emergency Plan**.

Even though this checklist looks long and scary, it is easy to do and can help you make a plan. We suggest you and your family review this list, then read the *entire* Manual since there are many tips mentioned in various topics and Sections that could help develop your plan.

PLEASE make some time in your busy lives to prepare for a disaster... a few minutes now could possibly save a life when a disaster hits!

Remember - be aware... be prepared... and have a plan!

## LEARN ABOUT RISKS & EXISTING PLANS:

*(See Section 4 for phone numbers of State & Provincial Emergency Management and Red Cross offices - or check city/county white pages)*

[ ]  Find out which disasters could occur in your area.

[ ]  Ask how to prepare for each disaster... but read this Manual first!

[ ]  Ask how you will be warned of an emergency.

[ ]  Learn your community's evacuation routes.

[ ]  Ask about special assistance for elderly or disabled persons.

[ ]  Ask your workplace about Emergency Plans.

[ ]  Learn about emergency plans for your children's school(s) or day care center(s).

## TIPS ON MAKING YOUR FAMILY PLAN:

*(Review all and complete Family Emergency Plan on pages 16-17)*:

[ ]  Meet with household members to talk about the dangers of fire, severe weather, earthquakes and other emergencies. Explain how to respond to each using the tips in this Manual.

[ ]  Find the safe spots in your home for each type of disaster. *(see Section 2 for explanations of each disaster)*

[ ] Talk about what to do when there are power outages and injuries.

[ ] Draw a floor plan of your home. Using a black or blue pen, show location of doors, windows, stairways and large furniture. Mark locations of emergency supplies, disaster kits, fire extinguishers, smoke detectors, collapsible ladders, first aid kits and utility shut-off points. Next, use a colored pen to draw a broken line charting at least two escape routes from each room.

[ ] Show family members how to turn off the water, gas and electricity at the main switches when necessary.

[ ] Post emergency telephone numbers near telephones.

[ ] Teach children how and when to call 9-1-1, police and fire departments *(see Section 3)*.

[ ] Make sure household members understand they should turn on the radio for emergency information.

[ ] Pick one out-of-state and one local friend or relative for family members to call if separated during a disaster. (It is often easier to call out-of-state than within the affected area.)

[ ] Pick two emergency meeting places in case you can't go home.
    1. A place near your home.
    2. A place outside the neighborhood

[ ] Teach children emergency phone numbers and meeting places.

[ ] Take a basic first aid and CPR class. *(See Section 3 for some Red Cross programs)*

[ ] Practice emergency evacuation drills with all household members at least two times each year.

[ ] Keep family records in a water- and fire-proof container. Consider keeping another set of records in a safety deposit box offsite.

[ ] Check if you have enough insurance coverage. *(See Section 2 for more information on flood insurance.)*


## TIPS FOR ELDERLY & DISABLED FAMILY MEMBERS:

[ ] Ask about special aid that may be available in an emergency for elderly and disabled family members. Find out if assistance is available for evacuation and in public shelters. Many communities

ask people with a disability to register with local fire departments or emergency management office so help can be provided quickly in an emergency. Check if this is available in your community!!

[ ] Ask children's teachers and caregivers about emergency plans for schools, day care centers or nursing homes.

[ ] If you currently have a personal care attendant from an agency, check to see if the agency will be providing services at another location if there is an evacuation -- and tell family members.

[ ] Learn what to do for each type of emergency. For example, basements are not wheelchair-accessible so you should have alternate safe places for different types of disasters for disabled or elderly persons.

[ ] Learn what to do in case of power outages and personal injuries. Know how to connect or start a back-up power supply for essential medical equipment!

[ ] If someone in the home uses a wheelchair, make sure 2 exits are wheelchair-accessible in case one exit is blocked.

[ ] Consider getting a medical alert system that will allow you to call for help if you have trouble getting around.

[ ] Both elderly and disabled persons should wear a medical alert bracelet or necklace at all times if they have special needs.

[ ] Consider setting up a "Buddy" system with a roommate, neighbor or friend. Give this person a copy of your **Family Emergency Plan** phone numbers and keep them updated of any changes. Give "buddy" an extra house key or tell them where one is available.

[ ] Consider putting a few personal items in a lightweight drawstring bag (e.g. a whistle, some medications, a small flashlight, extra hearing aid batteries, etc.) and tie it to your wheelchair or walker for emergencies. Make sure to rotate items so current and working.

[ ] Visit the **National Organization on Disability** web site to learn more about Emergency Preparedness issues at www.nod.org

## TIPS FOR PETS OR LIVESTOCK/LARGE ANIMALS:

### TIPS FOR PETS

[ ] If you have to evacuate your home, DO NOT leave pets behind! Make sure you have a secure pet carrier, leash or harness so if it panics, it can't run away.

[ ] For public health reasons, many emergency shelters cannot accept pets (unless it is a service animal assisting a disabled person). Find out which motels and hotels in your area allow pets in advance of needing them. Include your local animal shelter's number on next page since they might provide information during a disaster.

[ ] Make sure identification tags are up to date and securely fastened to your pet's collar. Keep a photo handy in wallet for identification purposes - just in case!

[ ] Make sure a roommate, neighbor or friend has an extra house key to evacuate your pets in the event you are unavailable.

## TIPS FOR LIVESTOCK/LARGE ANIMALS

[ ] Evacuate livestock whenever possible. Make arrangements for evacuation, including routes and host sites, in advance. Alternate routes should be mapped out as a backup.

[ ] The evacuation site should have food, water, veterinary care, handling, equipment and facilities.

[ ] Trucks, trailers, and vehicles for transporting animals should be available with experienced handlers and drivers to transport them.

[ ] If evacuation is not possible, a decision must be made whether to move large animals to available shelter or turn them outside. This decision should be based on the type of disaster and the soundness and location of the shelter or structure.

[ ] If you board animals, ask if facility has an evacuation plan in place.


Next, we suggest you sit with your family and write down part of your **Family Emergency Plan** using the next 2 pages as a guide. Put this information on sheets of paper near telephones where everyone can see it and keep it updated. Then review how to put together a **Disaster Supplies Kit** since you may not have much time if you are told to leave during a disaster or emergency.

Again, we suggest you and your family read this *entire* Manual together - especially your kids - since there are many tips here that could help you make a plan and learn what to do if the unexpected happens.

You may just want to review the book first and then come back to this Section later!

# FAMILY EMERGENCY PLAN

## EMERGENCY CONTACT NUMBERS

*(Post a copy of this information near each phone for easy access!)*

**Out-of-State Contact**
Name _____
City _____
Telephone (Day) _____ (Evening) _____

**Local Contact**
Name _____
Telephone (Day) _____ (Evening) _____

**Nearest Relative**
Name _____
City _____
Telephone (Day) _____ (Evening) _____

**Family Work Numbers**
Father _____ Mother _____
Guardian _____
Brother _____ Sister _____

**Emergency Telephone Numbers**
In a life-threatening emergency, dial 911 or local emergency medical
services system number

Police Department _____
Fire Department _____
Ambulance _____
Hospital _____
Poison Control  1-800-222-1222 (U.S. only) _____

**Family Doctors**
Name_____ Phone # _____
Name_____ Phone # _____

## EMERGENCY PLAN, continued

**Veterinarian:** _____

Animal Shelter or Humane Society: _____


In case you get separated from family members during an emergency or disaster, please decide on TWO Meeting Places or Areas where you can join each other.

Please make sure your small children are included when making this decision and they understand why they should meet here.


### Meeting Place or Meeting Area

1. Right outside your home _____

   _____

   *(Example: meet by the curb or by the mailbox in front of home or apartment building)*


2. Away from the neighborhood, in case you cannot return home

   _____

   *(Example: choose the home of a family friend or relative and fill in below)*

   Address _____

   Telephone # _____

   Directions to this place _____

   _____

   _____

# DISASTER SUPPLIES KIT

Disasters happen anytime and anywhere -- and, when disaster strikes, you may not have much time to respond. And sometimes services may be cut off or first responders can't reach people right away. Would you and your family be prepared to cope until help arrives?

Both the Red Cross and FEMA recommend keeping enough supplies in the home to meet your family's needs for at *least* three days or longer (up to 2 weeks or more, if possible). Once disaster threatens or hits, you may not have time to shop or search for supplies ... BUT, if you've gathered supplies in advance in your **Disaster Supplies Kit**, your family could handle an evacuation or shelter living easier. And since everything is all together in one place... all you gotta do is **GRAB & GO**!

Put items you'd most likely need (water, food, first aid, emergency items, etc.) in a container that is easy-to-carry and that will fit in your vehicle. For example, a large trash can or storage container with a lid that snaps shut tightly (some even come with wheels), or a waterproof backpack or large duffel bag (waterproof, if possible) would be useful.

We're also including suggestions for a CAR KIT and a CLASSROOM or LOCKER or OFFICE KIT since these are usually the most common places you would be if and when a disaster strikes.

There are seven basic categories of supplies you should stock in your home kit: **water**, **food**, **first aid supplies**, **tools and emergency supplies**, **sanitation, clothing and bedding**, and **special items**.

Take advantage of sales and stock up as you can -- also put dates on food cans or labels to show when they were purchased. Supplies should <u>ALL</u> be checked every 6 months to make sure they are still good and working! We suggest you mark dates on your calendar and have the entire family help check all the items together. Again, it will be good quality time with the family and will give you all a chance to update any phone numbers or information that has changed.

## WATER

A normally active person needs to drink at least 2 quarts (2 litres) of water each day and possibly as much as a gallon (4 litres) a day.

[ ] Store one gallon of water per person per day (two quarts/litres for drinking and two quarts/litres for food preparation and sanitation).

[ ] Keep at <u>least</u> a three-day supply of water for each person in your household. <u>Rotate</u> new bottles every 6 months.

[ ] Store extra bottles for pets so you don't reduce your amount.

[ ] Review TIPS ON WATER PURIFICATION at end of section 2.


## FOOD

Choose foods that require no refrigeration, preparation or cooking and little or no water.  If you must heat food, pack a can of sterno or a small propane camping stove.  Select foods that are compact and lightweight and <u>rotate</u> food out every 6 months.

[ ] Ready-to-eat canned meats, fish, fruits, and vegetables (and put in a <u>manual can opener</u>!!)

[ ] Canned juices, milk, soups (if powder or cubes, store extra water)

[ ] Staples - sugar, salt, pepper

[ ] High energy foods - peanut butter, jelly, crackers, granola bars, trail mix, nuts, jerky, dried fruits, Emergency Food bars, etc.

[ ] Vitamins & herbs (e.g. a good multiple, Vitamins C & E, garlic pills [boosts immune], L-Tyrosine [amino acid for stress], etc.)

[ ] Foods for infants, elderly persons or persons on special diets

[ ] Foods for your pet (if necessary)

[ ] Comfort / stress foods - cookies, hard candy, suckers, sweetened cereal, instant coffee, tea bags

[ ] Some companies offer survival and long-term storage foods that are freeze dried and sold in months, 1-year, and 2-year supplies


## FIRST AID KITS

You should always be prepared and keep a First Aid Kit in your home <u>and</u> in every car and make sure everyone knows where kits are and how to use them.  And if you like the outdoors (hiking, biking, etc.) you should carry a small Kit in your fanny pack or backpack as a precaution.

There are many different sizes of First Aid Kits on the market that vary in price. You can also make your own kits using things that may already be in your home. Consider including the following items in a **waterproof** container or bag so you can be prepared for almost any type of emergency!

We realize there are a <u>lot</u> of items suggested here, but the more you prepare … the better off you and your family will be during a disaster.

## Items to include in First Aid Kit

- Ace bandage(s)
- Adhesive bandage strips in assorted sizes
- Adhesive tape
- Antibiotic ointment or gel
- Antiseptic towelettes
- Assorted sizes of safety pins
- Box of Baking soda
- Cleansing agent (isopropyl alcohol, hydrogen peroxide and/or soap)
- Cold pack
- Contact lens solution and Eyewash solution
- Cotton and Cotton swabs
- Copy of *IT'S A DISASTER!* manual
- Dental repair kit (usually near toothpaste section)
- Disposable Face shield for Rescue Breathing
- Disposable gloves
- Flashlight & batteries - check often to make sure it works & batteries are good (Tip: remove batteries while stored but keep together)
- Gauze pads
- Heat pack
- Hydrogen peroxide
- Lip balm (one with SPF is best)
- Moleskin (for blisters on feet)
- Needles
- Petroleum jelly or other lubricant
- Plastic bags
- Roller gauze
- Scissors
- Small bottle of hand lotion
- Snake bite kit with extractor
- Sunscreen (choose one between SPF 15 and SPF 30)
- Thermometer
- Triangular bandages
- Tweezers

## Non-prescription drugs to include in First Aid Kit

- Activated charcoal (use if advised by the Poison Control Center)

- Antacid (for upset stomach)
- Anti-diarrhea medication
- Antihistamine and decongestant (for allergic reactions or allergies and sinus problems)
- Aspirin, acetaminophen, ibuprofen and naproxen sodium
- Laxative
- Potassium Iodide  *(see NUCLEAR POWER PLANT EMERGENCY)*
- Syrup of ipecac (used only if advised by Poison Control Center)
- Vitamins & herbs (e.g. a good multiple, Vitamins C & E, garlic pills or zinc [boosts immune sys], L-Tyrosine [amino acid for stress], etc.)

## Prescription drugs to include in First Aid Kit

Since it may be hard to get prescriptions filled during a disaster, talk to your physician or pharmacist about storing these types of medications. And make sure to check labels for special instructions and expiration dates.

## Tips on First Aid items that are inexpensive and widely available:

**Activated charcoal** - absorbs poisons and drugs in the stomach and intestines and helps prevent toxins from being absorbed into the bloodstream by coating intestinal walls. (You should check with the Poison Control Center before taking since it doesn't work on all toxic substances.) It is found at natural foods stores and pharmacies in powder, liquid, and capsule forms. The capsules can also be broken open to use powder for making a paste on insect bites and stings.

**Baking soda** - aid for occasional heartburn or indigestion; use as substitute for toothpaste; sprinkle in bath water for sore muscles or bites & stings; or make a paste (3 parts baking soda to 1 part water) to use on bee stings or insect bites, poison ivy, canker sores, sunburn, and rashes (but is too strong for infants!)

**Hydrogen peroxide** - can help clean and disinfect wounds, treat canker sores, gingivitis, and minor earaches. Also can be used for cleaning hands or for brushing teeth. (The reason it foams up on skin or item is because of the oxygen at work - means it's killing germs!)

**Meat tenderizer** - (check ingredient list on bottle for "papain") make a paste to use on insect bites and stings. Papain is a natural enzyme derived from papaya that can help break down insect venom.

**Syrup of ipecac** [pronounced ip'- î - kak] - use only when advised by the Poison Control Center to induce vomiting (makes you puke) -- available at most pharmacies or drug stores in 1 oz bottles.

**Vinegar** - helps relieve jellyfish stings, sunburn, and swimmer's ear.

# TOOLS AND EMERGENCY SUPPLIES

Items that may come in handy if you have to evacuate or if stuck at home without power.

[ ] Aluminum foil and resealable plastic bags

[ ] Battery-operated radio and extra batteries (remember to check batteries every 6 months). Also consider radios like the NOAA Weather Radio and Environment Canada's Weatheradio with one-alert feature that automatically alerts you when a Watch or Warning has been issued.

[ ] Battery-operated travel alarm clock

[ ] Cash or traveler's check and some change

[ ] CD-Rom (can be used as a reflector to signal planes if stranded)

[ ] Compass

[ ] Extra copy of *IT'S A DISASTER!* manual

[ ] Flashlight and extra batteries & extra bulbs (check every 6 months)

[ ] Fire extinguisher: small canister, ABC type

[ ] Manual can opener and a utility knife

[ ] Map of the area (to help locate shelters)

[ ] Matches in a waterproof container and candles

[ ] Medicine dropper (e.g. measure bleach to purify water, etc.)

[ ] Needles & thread

[ ] Paper, pencil (store in baggies to keep dry)

[ ] Paper cups, plates, plastic utensils (or Mess Kits) and paper towels

[ ] Plastic sheeting (for shelter, lean-to, or sealing room during chemical / hazardous material alert - *see HAZARDOUS MATERIALS*)

[ ] Pliers

[ ] Tape (plastic & duct)

[ ] Signal flare

[ ] Small shovel or trowel

[ ] Sterno or small camp stove and mini propane bottle

[ ] Wrench (to turn off household gas and water)

[ ] Whistle (can be used to call for help in an emergency)

[ ] Work gloves

## SANITATION

Make sure all these items are in a waterproof containers or plastic bags.

[ ] Disinfectant *(See TIPS ON SANITATION OF HUMAN WASTE)*

[ ] Feminine supplies  (tampons, pads, etc.)

[ ] Household chlorine bleach (regular scent)

[ ] Personal hygiene items (toothbrushes, toothpaste or baking soda, brush, comb, deodorant, shaving cream, razors, etc.)

[ ] Plastic garbage bags, ties (for personal sanitation uses)

[ ] Plastic bucket with tight lid (for human waste use)

[ ] Soap, liquid detergent, waterless hand sanitizer, hydrogen peroxide

[ ] Toilet paper, baby wipes

[ ] Wash cloths, hand and bath towels

## CLOTHING AND BEDDING

[ ] At least one complete change of clothing and footwear per person

[ ] Sturdy shoes or work boots and extra socks

[ ] Hats, gloves and thermal underwear

[ ] Blankets or sleeping bags (small emergency ones are cheap and about the size of a wallet ... or pack extra garbage bags)

[ ] Rain gear or poncho (small emergency ones are cheap and about the size of a wallet or use plastic garbage bags)

[ ] Safety glasses and/or Sunglasses

[ ] Small stuffed animal, toy or book for each child at bedtime

## SPECIAL ITEMS

[ ] Entertainment -  games, books and playing cards

[ ] Important Family Documents  (keep in waterproof, portable safe container and update when necessary!)
— Extra set of car keys, cash, traveler's checks and credit card
— Will, insurance policies, contracts, deeds, stocks and bonds
— Passports, social security cards, immunization records
— Bank account numbers
— Credit card numbers and companies
— Inventory of valuable household goods + phone numbers
— Family records (birth, marriage, death certificates)
— Recent pictures of all family members and pets for identification needs

[ ] RED and GREEN construction paper or RED and GREEN crayons or markers (can signal rescue workers to stop or move on)


Remember to pack things for family members with special needs such as Infants, Elderly and Disabled persons, and Pets:

For Infants

[ ] Bottles

[ ] Diapers, baby wipes and diaper rash ointment

[ ] Formula and cereals

[ ] Medications

[ ] Powdered milk and juices

[ ] Small soft toys


For Elderly and Disabled (Children & Adults)

[ ] Bladder control garments and pads

[ ] Denture needs

[ ] Extra eye glasses or contact lenses and supplies

[ ] Extra hearing aid batteries

[ ] Extra wheelchair batteries, oxygen, catheters or other special equipment

[ ] A list of style and serial numbers of medical devices such as pacemakers, etc. and copy of Medicare card

[ ] List of prescription medications and dosages or allergies (if any)

[ ] Special medicines for heart, high blood pressure, diabetes, etc.

[ ] Store backup equipment (such as a manual wheelchair, cane or walker) at a neighbor's home or at another location

For Pets

[ ] Cage or carrier, bedding, leash, muzzle, kitty litter, etc.

[ ] Chew toys or treats

[ ] Medications or special foods

# CAR KIT

Keep most or all of these items in a waterproof pack so everything is together and easy to grab. Make one for each vehicle too!

[ ] Battery-powered radio, flashlight, extra batteries and extra bulbs

[ ] Blanket (small emergency ones are cheap and size of a wallet)

[ ] Bottled water and non-perishable foods (Tip: store food in empty coffee cans to keep it from getting squashed)

[ ] CD-Rom (can be used as a reflector to signal planes if stranded)

[ ] Copy of *IT'S A DISASTER!* manual

[ ] Extra clothes (jeans and sweater), sturdy shoes and socks

[ ] First Aid Kit

[ ] Local maps

[ ] Plastic bags that seal

[ ] Shovel (small collapsible ones are available)

[ ] Short rubber hose (for siphoning)

[ ] Small fire extinguisher (5 lb., ABC type)

[ ] Tools - Tire repair kit, booster cables, flares, screw driver, pliers, knife, wire

[ ] Work gloves

## CLASSROOM OR LOCKER OR OFFICE KIT

Keep items in a small pack, drawstring bag or duffel so everything is together and easy to grab!

[ ] Battery-operated radio and extra batteries

[ ] Emergency blanket (small, cheap, & light - size of a wallet)

[ ] Extra copy of *IT'S A DISASTER!* manual

[ ] A few plastic trash bags

[ ] Mini or regular flashlight and extra bulbs and batteries

[ ] Non-perishable foods like crackers, cookies, trail mix, granola bars, etc. (Ask children to help choosing food and make sure they understand this is for Emergencies!)

[ ] Small (plastic) bottled water or juice… as much as you can fit

[ ] Small First Aid kit

[ ] Small stuffed animal, book, or toy for children

[ ] Small packet of tissues

[ ] Small packet of moist towelettes or mini bottle of hand sanitizer (waterless kind)

[ ] Sweatshirt or sweater

[ ] Work gloves to protect your hands (especially from broken glass)

# Suggestions & Reminders About Kits

- Store your **Disaster Supplies Kit** in convenient place known to <u>ALL</u> family members. Keep a smaller version in the trunk / back of every vehicle (see CAR KIT).

- Keep items in airtight plastic bags to keep them dry in kit.

- Take advantage of end-of-season clearance sales and grocery sales (esp. can goods) and stock up as you can. Look around your home since you may be able to put a lot of these things together from what is already on shelves or in drawers or medicine cabinets.

- Replace your stored food and water supply every 6 months! It's best to test or replace batteries at this time too. Make a game of it by keeping track on a calendar or on a poster drawn by children so they can help. Also, everyone should meet every 6 months anyway to go over the **Family Emergency Plan** and update any data (phone numbers, address changes, etc.)

- Ask your physician or pharmacist about storing prescription medicines.

- Visit the U.S. Department of Homeland Security web site for more tips about Kits at <u>www.ready.gov</u>

# Section 2

## Disaster Preparedness & Prevention

# What to Do **BEFORE** A Disaster Strikes (Mitigation Tips)...

There are many things you can do to protect yourself, your home and your property BEFORE any type of natural hazard or disaster strikes. One of the most important things citizens can do is learn about hazards and risks in your area and take personal responsibility to prepare for the unexpected.

Please realize that natural disasters have common elements that overlap (like wind and floods) and we are only summarizing some key topics here to help get you started.

There are many mitigation tips and programs available from government agencies, public and private businesses, nonprofits and NGOs listed here and in **Section 4** of this book that can help you and your family learn more.

## What is Mitigation?

Mitigation simply means an effort to lessen the impact disasters have on people, property, communities and the economy. It is also about reducing or eliminating risks before disasters strike and involves planning, commitment, preparation and communication between local, state and federal government officials, businesses and the general public.

Some examples of mitigation include installing hurricane straps to secure a structure's roof to its walls and foundation, building outside of flood plains, securing shelves and loose objects inside and around the home, developing and enforcing effective building codes and standards, using fire-retardant materials ... and the list goes on and on.

Soon we will explain what to do BEFORE, DURING and AFTER specific types of natural and man-made disasters and emergencies. But first there are some things you should do in advance that take time and planning... otherwise known as prevention or mitigation tips.

First we'll cover some mitigation strategies available for businesses and consumers, then cover mitigation tips on the two most common disasters (**winds** and **floods**) followed by other topics listed alphabetically. Also please review the BEFORE sections on common disasters that occur in your area since there may be a few other mitigation tips there too.

Remember... the more you prepare BEFORE disaster strikes, the better off you and your loved ones will be financially, emotionally and physically!

## Mitigation Strategies for Businesses & Consumers

Both the United States and Canada have national programs designed to help the public, businesses and communities prepare for the unexpected.

In the U.S., the Federal Emergency Management Agency (FEMA) merged the Federal Insurance Administration and the Mitigation Directorate to create the Federal Insurance and Mitigation Administration (FIMA). FIMA combines organizational activities to promote Protection, Prevention, and Partnerships at the Federal, State, Local and individual levels to lessen the impact of disasters upon families, homes, communities and economy through damage prevention and flood insurance.

FIMA is made up of a number of programs and activities like the National Flood Insurance Program, National Hurricane, National Dam Safety, and National Earthquake Hazards Reduction Programs, and others involving Mitigation. For example, Mitigation Grant Programs provide funding for State and Local governments to reduce the loss of lives and property on future disasters, and Mitigation Planning offers resources to determine risks and hazards in communities. Plus FIMA provides citizens information about "safe rooms" and flood insurance, and small businesses can learn about Pre-Disaster Mitigation (PDM) loans and other cost-saving mitigation tips for structures and property.

*To learn more please visit FIMA online at www.fema.gov/fima*

In Canada, the Office of Critical Infrastructure Protection and Emergency Preparedness (OCIPEP) is working with federal departments and agencies to determine how the Government of Canada can support the development of a National Disaster Mitigation Strategy and the co-operative arrangements that are needed for its implementation. Consultations with provincial / territorial governments, non-governmental organizations, and the private sector were held throughout 2002 with results to be summarized into a proposed NDMS framework for further consideration in 2003.

*To learn more please visit OCIPEP online at www.ocipep.gc.ca*

## MITIGATION TIPS TO HELP PREVENT DAMAGE AND LOSS:

### WIND MITIGATION *(MOST COMMON)*
Wind damage is the most common disaster-related expense and usually accounts for about 70% or more of the insured losses reported worldwide. Many natural disasters like hurricanes, tornadoes, microbursts or thunderstorms, and winter storms include damaging winds. And certain parts of the world experience high winds on a normal basis due to wind patterns.

Realize when extreme winds strike they are not constant - they rapidly increase and decrease. A home in the path of wind causes the wind to change direction. This change in wind direction increases pressure on parts of the house creating stress which causes the connections between building components to fail. For example, the roof or siding can be pulled off or the windows can be pushed in.

## Strengthen weak spots on home

Experts believe there are four areas of your home that should be checked for weakness -- the roof, windows, doors and garage doors. Homeowners can take some steps to secure and strengthen these areas but some things should be done by an experienced builder or contractor.

ROOF:
- Truss bracing or gable end bracing (supports placed strategically to strengthen the roof)
- Anchors, clips and straps can be installed (may want to call a professional since sometimes difficult to install)

WINDOWS and DOORS:
- Storm shutters (available for windows, French doors, sliding glass doors, and skylights) or keep plywood on hand
- Reinforced bolt kits for doors

GARAGE DOORS:
- Certain parts of the country have building codes requiring garage doors to withstand high winds (check with local building officials)
- Some garage doors can be strengthened with retrofit kits (involves installing horizontal bracing onto each panel)

## Secure mobile homes

Make sure your trailer or mobile home is securely anchored. Consult the manufacturer for information on secure tiedown systems.

## Secure or tie down loose stuff

Extreme winds can also cause damage from flying debris that can act like missiles and ram through walls, windows or the roof if the wind speeds are high enough. You should consider securing large or heavy equipment inside and out to reduce some of the flying debris like patio furniture, barbeque grills, water heaters, garbage cans, bookcases and shelving, etc.

## Consider building a shelter or "safe room"

Shelters or "safe rooms" are designed to provide protection from the high winds expected during hurricanes, tornadoes and from flying debris. Shelters built below ground provide the best protection, but be aware they could be flooded during heavy rains.

FEMA provides an excellent free booklet called "Taking Shelter From the Storm: Building a Safe Room Inside Your House" developed in association with the Wind Engineering Research Center at Texas Tech University.

*Learn more about safe rooms by visiting www.fema.gov/mit/saferoom*

## FLOOD MITIGATION (2ND MOST COMMON)

Flood damage is normally the second most common disaster-related expense of insured losses reported worldwide. Many natural disasters like hurricanes, tornadoes, rain, thunderstorms, and melting snow and ice cause flooding.

There are certain parts of North America known as "flood plains" that are at high risk of floods. You may want to contact your local emergency management official to develop a community-based approach and there may even be funds available to assist you and your area.

Some examples of State grant programs officials can access include the Hazard Mitigation Grant Program (**HMGP**), Flood Mitigation Assistance (**FMA**) Program, and the Pre-Disaster Mitigation (**PDM**) Program. Individual citizens cannot apply for grant money but local agencies or non-profit organizations may apply on behalf of citizens.

### But I have insurance...

Insurance companies will cover some claims due to water damage like a broken water main or a washing machine that goes berserk. However, standard home insurance policies DO NOT generally cover flood damage caused by natural events or disasters!

The United States offers a **National Flood Insurance Program** available in most communities and there is a waiting period for coverage. Talk to your local insurance agent, check the Yellow Pages directory, or contact NFIP at 1-888-379-9531 or 1-800-427-5593 (TTY) or visit www.floodsmart.gov

Currently Canadians do not have a national flood program, however, there are certain parts of Canada that offer limited flood-damage coverage but it must be purchased year-round and the rates are relatively high. The Insurance Bureau of Canada suggests you consult your insurance representative with questions regarding coverage.

### Move valuables to higher ground

If your home or business is prone to flooding, you should move valuables and appliances out of the basement or ground level floors.

### Elevate breakers, fuse box and meters

Consider phoning a professional to elevate the main breaker or fuse box and utility meters above the anticipated flood level so flood waters won't damage your utilities. Also consider putting heating, ventilation and air conditioning units in the upper story or attic to protect from flooding.

### Protect your property

Build barriers and landscape around homes or buildings to stop or reduce floodwaters and mud from entering (see pages 45-46). Also consider sealing basement walls with waterproofing compounds and installing "check valves" in sewer traps to prevent flood water from backing up into your drains.

The next few pages cover some key mitigation tips on several types of disasters and topics (sorted alphabetically). After this mitigation section we will then cover specific natural and man-made disasters in more detail.

## AIR QUALITY MITIGATION

We want to briefly mention indoor air quality here since it affects so many people at home, school and work (especially children and the elderly). Poor air quality often results naturally from many environmental and weather-related factors. There are things people can do and kits available for testing home, work and school environments so please learn more about carbon monoxide, mold, and radon by visiting or calling the following groups ...

**EPA's Indoor Air Quality**: www.epa.gov/iaq or call 1-800-438-4318

**Center for Disease Control's National Center for Environmental Health Air Pollution & Respiratory Health**: www.cdc.gov/nceh/airpollution

**National Radon Info Line**: 1-800-SOS-Radon (1-800-767-7236)

## EARTHQUAKE MITIGATION

A lot of the ongoing research by scientists, engineers and emergency preparedness officials has resulted in improvements to building codes around the world. Proven design and construction techniques are available that help limit damage and injuries.

There are some things you can do to reduce risk in an earthquake-prone area:

### Consider retrofitting your home

There are options to retrofit or reinforce your home's foundation and frame available from reputable contractors who follow strict building codes.

Other earthquake-safety measures include installing flexible gas lines and automatic gas shutoff valves. Changes to gas lines and plumbing in your house must be done by a licensed contractor who will ensure that the work is done correctly and according to code. This is important for your safety.

### Secure loose stuff

- Use nylon straps or L-braces to secure cabinets, bookcases and other tall furniture to the wall.
- Secure heavy appliances like water heaters, refrigerators, etc. using bands of perforated steel (also known as "plumber's tape").
- Use buckles or safety straps to secure computers, televisions, stereos and other equipment to tabletops.
- Use earthquake or florist putty to tack down glassware, heirlooms and figurines.

## FIRE MITIGATION

Home fire protection is very important and covered on pages 55-56. Also see Wildfire Mitigation below to learn additional ways to protect your home.

## LIGHTNING MITIGATION

Here are some safety tips to prepare your home for lightning.

### Install a Lightning Protection System

A lightning protection system does not prevent lightning from striking but does create a direct path for lightning to follow. Basically, a lightning protection system consists of air terminals (lightning rods) and associated fittings connected by heavy cables to grounding equipment. This provides a path for lightning current to travel safely to the ground.

### Install surge protectors on or in home

Surge protection devices (SPDs) can be installed in the electrical panel to protect your entire home from electrical surges. Sometimes it may be necessary to install small individual SPDs in addition to the home unit for computers and television sets due to different ratings and voltage levels.

If a home unit is too expensive, consider getting individual surge protection devices that plug into the wall for the refrigerator, microwave and garage door openers. Appliances that use two services (cable wire and electrical cord) may require combination SPDs for computers, TVs, and VCRs.

## WILDFIRE MITIGATION

As our population continues to grow, more and more people are building homes in places that were once pristine wilderness areas. Homeowners who build in remote and wooded areas must take responsibility for the way their buildings are constructed and the way they landscape around them.

### Use Fire Resistant Building Materials

The roof and exterior structure of your home and other buildings should be constructed of non-combustible or fire-resistant materials. If wood siding, cedar shakes or any other highly combustible materials are used, they should be treated with fire retardant chemicals.

### Landscape wisely

Plant fire-resistant shrubs and trees to minimize the spread of fire and space your landscaping so fire is not carried to your home or other surrounding vegetation. Remove vines from the walls of your home.

### Create a "safety zone" around the house
- Mow grass regularly.

- Stack firewood at least 100 feet (30 m) away and uphill from home.
- Keep your roof and gutters free of pine needles, leaves, and branches and clear away flammable vegetation at least 30 to 100 feet (9 to 30 m) from around your structures.
- Thin a 15-foot (4.5 m) space between tree crowns and remove limbs within 10-15 feet (3 - 4.5 m) of the ground.
- Remove dead branches that extend over the roof.
- Prune tree branches and shrubs within 10 feet (3 m) of a stovepipe or chimney outlet.
- Remove leaves and rubbish from under structures.
- Ask the power company to clear branches from power lines.
- Keep combustibles away from structures and clear a 10-foot (3 m) area around propane tanks, barbeques, boats, etc.

## Protect your home
- Install smoke detectors, test them each month and change batteries once a year.
- Consider installing protective shutters or fire-resistant drapes.
- Inspect chimneys at least twice a year and clean every year.
- Cover chimney and stovepipe flue openings with 1/2 inch (1 cm) or smaller non-flammable mesh screen.
- Use this same mesh screen beneath porches, decks, floor areas and the home itself.  Also screen openings to attic and roof.
- Soak ashes and charcoal briquettes in water for two days in a metal bucket.
- Keep a garden hose connected to an outlet.
- Have fire tools handy (ladder, shovel, rake, saw, ax, bucket, etc.)
- Address should be visible on all structures and seen from road.


## WINTER STORM & EXTREME COLD MITIGATION
Severe winter weather causes deterioration and damage to homes every year. There are many things you can do to prepare for the bitter cold, ice and snow in advance to save you money and headaches in the long run.  Some of these tips should be used by apartment dwellers too!

## "Winterize" your home
- Insulate walls and attic.
- Caulk and weather-strip doors and windows to keep cold out.
- Install storm windows or cover windows with plastic film from the inside to keep warmth in.
- Detach garden hoses and shut-off water supply to those faucets.

- Install faucet covers or wrap tightly with towels and duct tape.
- Show family members the location of your main water valve and mark it so you can find it quickly.
- Drain sprinkler water lines or well lines before the first freeze.
- Keep the inside temperature of your home at 68 degrees Fahrenheit (20 degrees Celsius) or higher, even if leaving.
- Wrap pipes near exterior walls with heating tape or towels.
- Change furnace filters regularly and have it serviced from time to time.
- Make sure you have good lighting from street and driveways to help others see snow and ice patches and try to keep paths clear of drifts.
- Remove dead tree branches since they break easily.
- Cover fireplace / stovepipe openings with fire-resistant screens.
- Check shingles to make sure they are in good shape.

## Preventing "ice dams"

A lot of water leakage and damage around outside walls and ceilings are actually due to "ice dams". Ice dams are lumps of ice that form on gutters or downspouts and prevent melting snow from running down. An attic with no insulation (like a detached garage) or a well-sealed and insulated attic will generally not have ice dams. But if the roof has many peaks and valleys, is poorly insulated, or has a large roof overhang, ice dams usually happen.

Some tips to prevent ice dams:
- Keep gutters and downspouts clear of leaves and debris.
- Find areas of heat loss in attic and insulate it properly.
- Wrap or insulate heating duct work to reduce heat loss.
- Remove snow buildup on roof and gutters using snow rake or soft broom.
- Consider installing roof heat tapes (electric cables) that clip onto the edge of your shingles to melt channels in the ice. (Just remember - cables use a lot of energy and may not be pretty but could help on older homes with complicated roofs).

## Preventing frozen pipes
- Keep cabinet doors open under sinks so heat can circulate.
- Run a slow trickle of lukewarm water and check water flow before going to bed and when you get up. (The first sign of freezing is reduced water flow so keep an eye on it!)
- Heat your basement or at least insulate it well!
- Close windows and keep drafts away from pipes since air flow can cause pipes to freeze more often.

# MITIGATION TIPS SUMMARY...

### Take responsibility...

Basically, no matter where you live, you should take personal responsibility and prepare yourself, your family and your property BEFORE disasters or natural hazards strike.

### ...and learn more!

After reviewing the remainder of this manual, please contact your local emergency officials or your local building department to learn about all the risks in your area and what to expect if disaster strikes.

Remember, the best thing you can do to deal with ANY type of disaster is...

BE AWARE... BE PREPARED... and... HAVE A PLAN!

If you do these 3 things, the life and property you save could be your own... because what you <u>don't</u> know <u>CAN</u> hurt you!

# What to Expect When **ANY** Type of Disaster Strikes...

Local government and disaster-relief organizations will try to help you but there are <u>many</u> times they cannot reach you immediately after a disaster.

You should be ready to be self-sufficient for at *least* three days... possibly longer depending on the type of disaster!

This may mean providing for your own shelter, food, water and sanitation. If you have planned ahead, it will be easier to recover from a disaster as long as you have a **Family Emergency Plan** and a **Disaster Supplies Kit** for you and your family. This can help reduce some of the fear, anxiety and losses that surround a disaster.

By planning ahead, you will know where to go, be ready to evacuate if necessary, and be a little more comfortable in a shelter by having some of your own personal items with you in your **Disaster Supplies Kit**.

Now we are going to explain what to do **BEFORE**, **DURING** and **AFTER** specific types of natural and man-made disasters (sorted alphabetically).

Then we'll cover some tips on **RECOVERING FROM A DISASTER** (includes many "AFTER" tips that apply to most every type of disaster) and on **SHELTER LIVING**.

We then offer some tips on **USING HOUSEHOLD FOODS, WATER PURIFICATION**, and **SANITATION OF HUMAN WASTE** followed by tips for **HELPING OTHERS** at the end this Section.

**Section 3** covers a variety of basic First Aid topics that may be necessary to use during a major disaster, emergency or just for the minor injury at home.

**Section 4** contains many helpful telephone numbers of organizations in America and Canada. And finally, we ask you please take some time to review the topics, resources and web sites near the back of this manual.

As we mentioned in the Introduction, a majority of this information was compiled from various publications provided by the American and Canadian Red Cross, U.S.'s Department of Homeland Security and FEMA, Canada's PSEPC and others to help assist you in preparing for various types disasters.

We realize you may not experience every type of disaster or emergency in your part of the world but, if you ever travel away from home, you could potentially be placed in a disaster situation so please educate yourself and your family. Knowledge is power and can help reduce fear and anxiety.

# What are <u>YOU</u> gonna do about...

# What are <u>YOU</u> gonna do about…
## AVALANCHES, LANDSLIDES & MUDFLOWS?

**Avalanches** - masses of loosened snow or ice that tumble down the side of a mountain, often growing as it descends picking up mud, rocks, trees and debris triggered by various means including wind, rapid warming, snow conditions and humans.

**Landslides** - masses of rock, earth or debris that move down a slope and can be caused by earthquakes, volcanic eruptions, and by humans who develop on land that is unstable.

**Mudflows** - rivers of rock, earth, and other debris soaked with water mostly caused by melting snow or heavy rains and create a "slurry". A "slurry" can travel several miles from its source and grows in size as it picks up trees, cars, and other things along the way just like an avalanche!

*Please realize data on avalanches fill up entire books and we are briefly touching on some basic information here with some references to obtain more information, then we'll cover landslides and mudflows.*

## Avalanche Basics

Snow avalanches are a natural process and happen about a million times per year worldwide. Contrary to what is shown in the movies, avalanches are <u>not</u> triggered by loud noises like a shout or a sonic boom -- it's just not enough force. An avalanche is actually formed by a combination of several things -- a steep slope (the terrain), the snowpack, a weak layer in the snow-pack, and a natural or artificial "trigger".

Nearly all avalanches that involve people are triggered by the victim or a member of their party. Each year avalanches claim between 100-200 lives around the world and thousands of people are partly buried or injured in them.

Millions of skiers, hikers, climbers, boarders, and snowmobilers venture out to enjoy winter sports each year pushing towns, roads and activities into avalanche-prone areas. Compound that with recreationists who cross into the backcountry with little or no basic avalanche training... and you've got a recipe for potential disaster!

## Types of avalanches

<u>Slab</u> - the most dangerous type of avalanche since it causes most fatalities. Experts compare slab avalanches to a dinner plate sliding off the table - a heavier plate of snow slides on top of weaker snow down a slope. An average-sized dry slab avalanche travels about 80 mph (128 km/h) and it's nearly impossible to outrun it or get out of the way!

Most avalanche deaths are caused by slab even though there are many obvious signs that indicate danger -- so educate yourself before venturing out into the backcountry!

Powder or loose snow - fresh fallen, light, dry snow (similar to fine sugar) rolls downhill with speeds of 110-180 mph (180-290 km/h) and swirls of powder climbing several thousand feet into the air. This is the most common type of avalanche and the danger is usually not the weight or volume but rather victims being pushed over a cliff or into a tree.

Some other types of avalanches include ice falls, wet and point release.

*You can find more information on the Internet at the WestWide Avalanche Network web site www.avalanche.org or visit your local library.*

## Typical Avalanche Victims

Nearly everyone caught in an avalanche is either skiing, snowboarding, riding a snowmobile, snowshoeing, hiking or climbing in the backcountry and they, or someone in their party, almost always trigger the avalanche that injures or kills them. According to the American Avalanche Association, the majority of victims are white, educated men between the ages of 18-35 who are very skilled at their sport.

One key is for the public to take personal responsibility and learn more about avalanche risks and safety procedures. The AAA has seen an increase in attendance now that avalanche educators are re-designing their courses to accommodate snowmobilers, snowboarders and other groups.

People should be prepared and learn how to recognize, assess and avoid avalanche danger by taking an avalanche-related course before entering the backcountry.

## The "Avalanche Triangle"

The following was excerpted from the **U.S.D.A. Forest Service National Avalanche Center** at www.fsavalanche.org (see "Avalanche Basics"):

Avalanches are formed by a combination of 3 ingredients (sometimes called the "avalanche triangle")...

Terrain - the slope must be steeper than 25 degrees and most often occur on slopes between 35 and 45 degrees. Most slab avalanches occur on slopes with starting zone angles between about 30 and 45 degrees.

Snowpack - the snowpack accumulates layer by layer with each weather event and both strong and weak layers exist. Strong layers contain small round snow grains that are packed closely together and well bonded (or cohesive). Weak layers are less dense and appear loose or "sugary". When a strong dense layer is over a weak less dense layer it's like a brick on top of potato chips -- the chips can't hold up the weight of the brick so an avalanche

occurs. Backcountry recreationists must learn the relationship of these  layers because weak layers prevent strong layers from bonding with one another thus causing unstable conditions.

A snowpack is balanced between stress and strength -- add additional stress
(like more snow or a human) and an avalanche could be triggered.

Weather - precipitation, wind and temperature can alter the stability of the
snowpack by changing the balance between stress and strength.  The type of
precipitation and at what rate it falls are equally as important as the amount.
If a lot of snow falls in a short amount of time, the snowpack has less time to
adjust to the additional stress.  Wind can blow large amounts of snow around
shifting the stress on the snowpack.  And rapid warming temperatures can
cause snowpacks to become very wet and unstable.

## BEFORE AN AVALANCHE:

Learn risks - Ask about local risks by contacting your local emergency management office *(see Section 4 for State & Provincial listings),* especially if
visiting or moving to an "avalanche-prone" area.

Take a course - Professional trainers and educators offer a variety of avalanche safety training courses and levels ranging from recreational novices to
backcountry experts. To learn more visit www.avalanche.org and click on
"Resources" then "Education".

Know your colors - Learn the Avalanche Danger Scales and corresponding
colors used where you live or plan to visit.

Get equipped - Carry avalanche rescue equipment or gear like portable
shovels, collapsible probes or ski-pole probes, high frequency avalanche
beacons (transceivers), etc. and learn how to use it!  Remember ... just
having avalanche equipment will NOT keep you out of an avalanche!!

Check it out - Check forecasts and avalanche advisories before going out.

Turn it on - Switch beacon on prior to entering the backcountry!  Check the
battery strength and verify the "transmit" and "receive" functionality with
everyone in your group to ensure beacons are picking up both signals.

Secure it - Before crossing a snow covered slope in avalanche terrain, fasten
clothing securely to keep snow out and remove your ski pole straps.

## DURING AN AVALANCHE:

Bail - Try out get out of the way if possible!  (For example, if you are a skier
or boarder - ski out diagonally... if on a snowmobile - drive downhill, etc.)

### If YOU are caught in the avalanche...

Scream and drop it - Yell and drop your ski poles (or anything in your hands) so they don't drag you down.

Start swimming - Use "swimming" motions, thrusting upward to try to stay near the surface of the snow.

Prepare to make an air pocket - Try to keep your arms and hands moving so the instant the avalanche stops you can make an air pocket in front of your face by punching in the snow around you before it sets.

### If you see SOMEONE ELSE caught in the avalanche...

Watch - Keep an eye on victim as they are carried downhill, paying particular attention to the last point you saw them.


## AFTER AN AVALANCHE:

### If YOU are caught in the avalanche...

Make an air pocket ASAP! - The INSTANT the avalanche stops try to maintain an air pocket in front of your face by using your hands and arms to punch in the snow and make a pocket of air. (You only have 1-3 seconds before the snow sets -- and most deaths are due to suffocation!)

Stick it out - If you are lucky enough to be near the surface, try to stick out an arm or a leg so that rescuers can find you.

Don't panic - Keep your breathing steady to help preserve your air space and help your body conserve energy.

Listen for rescuers - Since snow is such a good insulator, rescuers probably won't even hear you until they are practically on top of you, so don't start yelling until you hear them. (This will conserve your precious air!)

### If you see SOMEONE ELSE caught in the avalanche...

Watch - Keep watching the victim(s) as they are carried downhill, paying particular attention to the last point you saw them.

DO NOT go for help! - Sounds crazy but the victim only has a few minutes to breathe under the snow, so every second counts! Spend 30 minutes to an hour searching before going for help (unless you have a large party and someone can go while the rest search).

Be aware - Assess the situation and dangers... in many cases it is safe to go in after the avalanche settles but proceed with caution!

<u>Look for clues</u> - Start looking for any signs on the surface (like poles, a hand or foot, etc.) where victim was last seen. And remember, equipment and clothing can be ripped off during the avalanche but can help determine the direction they were carried.

<u>Switch to "receive"</u> - Turn all transceivers to "receive" to try to locate victim's signal (in the event victim is wearing one and has it set correctly!)

<u>Mark the spot</u> - If you lost sight of the victim or can't find any visible clues on the surface, mark the spot where victim was last seen.

<u>Probe in a line</u> - When searching with probes, stand shoulder to shoulder in a line across the slope and repeatedly insert probes moving down the slope.

<u>Listen</u> - Make sure you listen for any muffled sounds as you search.

<u>Find them...dig 'em out!</u> - If you find the victim, dig them out as quickly as possible! Survival chances reduce the longer they are buried.

*To learn more about avalanches visit the WestWide Avalanche Network web site at <u>www.avalanche.org</u>. Or see our ADDITIONAL RESOURCES & WEB SITES listed at the end of this book.*


Now we will briefly cover **landslides** and **mudflows**. Realize many types of disasters like earthquakes, volcanic eruptions, rain and wind erosion can cause land, rocks and mud to shift and move, sometimes at rapid speeds. Compound that with gravity and these earth movements can become extremely destructive.

Another major factor is the world's growing population is sprawling out of major cities and developing in high-risk areas. There are some warning signs to indicate if you have a potential problem.

## BEFORE A LANDSLIDE OR MUDFLOW:

<u>Learn risks</u> - Ask your local emergency management office *(see Section 4 for State & Provincial listing)* if your property is a "landslide-prone" area. Or contact your County/Municipal or State/Provincial Geologist or Engineer.

<u>Get insurance...?</u> - Talk to your agent and find out more about the **National Flood Insurance Program** since mudflows are covered by NFIP's flood policy. *(see FLOOD MITIGATION at beginning of this Section)*

<u>Be prepared to evacuate</u> - Listen to local authorities and leave if you are told to evacuate. *(see EVACUATION)*

<u>Reduce risks</u> - Plant ground cover on slopes and build retaining walls.

Inspect - Look around home and property for landslide warning signs:

- cracks appear on hill slopes, ground or paved roads
- water or saturated ground in areas not normally wet
- evidence of slow, downhill movement of rock and soil
- tilted trees, poles, decks, patios, fences or walls
- visible changes like sags and bumps at base of a slope
- doors and windows stick or cracks appear on walls, etc.

Call an expert...? - Consult a professional landscaping expert for opinions and advice on landslide problems.

## DURING A LANDSLIDE OR MUDFLOW:

Strange sounds - Listen for trees cracking, rocks banging together or water flowing rapidly (especially if near a stream or river) - could be close by!

Move it! - Whether you are in a vehicle, outside, or in your home – GET TO SAFER GROUND!

Be small - If there is no way to escape, curl into a tight ball and protect your head the best you can.

*(Since most other disasters cause landslides and mudflows, we'll discuss them further in those specific cases - please see other topics to learn more.)*

## AFTER A LANDSLIDE OR MUDFLOW:

Listen - Local radio and TV reports will keep you posted on latest updates or check with your local police or fire departments.

Don't go there - Stay away from the area until authorities say all is clear since there could be more slides or flows.

Insurance - If your home suffers any damage, contact your insurance agent and keep all receipts for clean-up and repairs.

Things to watch for:

- **flooding** - usually occur after landslides or debris flows
- **damaged areas** - roadways and bridges may be buried, washed-out or weakened -- and water, gas & sewer lines may be broken
- **downed power lines** - report them to power company

Replant - Try to fix or replant damaged ground to reduce erosion.

# What are <u>YOU</u> gonna do about...
## AN EARTHQUAKE?

Earthquakes can cause buildings and bridges to collapse, down telephone and power lines, and result in fires, explosions and landslides. Earthquakes can also cause huge ocean waves, called tsunamis, which travel long distances over water until they hit coastal areas.

Our planet's surface is actually made up of slowly-moving sections (called "tectonic plates") that can build up friction or stress in the crust as they creep around. An earthquake occurs when this built up stress is suddenly released and transmitted to the surface of the earth by earthquake waves (called seismic waves).

There are actually about one million small earthquakes, or seismic tremors, per year around the world. Many earthquakes are too small to be felt, but when they happen, you will feel shaking, quickly followed by a rolling motion that can rotate up, down, and sideways that lasts from a few seconds to several minutes!

## BEFORE AN EARTHQUAKE:

<u>Learn the buzzwords</u> - Learn the terms / words used with earthquakes...

- **Earthquake** - a sudden slipping of the earth's crust that causes a series of vibrations
- **Aftershock** - usually not as strong as an earthquake but can occur for hours, days, months or years after the main quake
- **Fault** - area of weakness where two sections of crust have separated
- **Epicenter** - area of the earth's surface directly above the crust that caused the quake
- **Seismic Waves** - vibrations that travel from the center of the earthquake to the surface
- **Magnitude** - used to define how much energy was released (A Richter Scale is the device used to measure this energy on a scale from 0-10 ... each whole number equals an increase of about 30 times the energy released meaning a 5.0 is about 30 times stronger than a 4.0.)

<u>Prepare</u> - See EARTHQUAKE MITIGATION at beginning of this Section.

<u>Reduce risks</u> - Look for things that could be hazardous...

- Place large or heavy objects on lower shelves and fasten shelves to walls, if possible.

- Hang heavy pictures and mirrors away from beds.
- Store bottled foods, glass, china and other breakables on low shelves or in cabinets that can fasten shut.
- Repair any faulty electrical wiring and leaky gas connections.

Learn to shut off - Know where and how to shut off electricity, gas and water at main switches and valves -- ask local utilities for instructions.

Do drills - Hold earthquake drills with your family to learn what to do...
- **DUCK** - drop down to the floor
- **COVER** - get under a heavy desk or table or against inside wall and protect head & neck with your arms
- **HOLD** - grab something sturdy and be ready to move with it

Make a plan - Review Section 1 and develop a **Family Emergency Plan.**

Check policies - Review your insurance policies.  Some damage may be covered even without specific earthquake insurance.


## DURING AN EARTHQUAKE:

Stay calm and stay where you are!  Most injuries happen when people are hit by falling objects when running IN or OUT of buildings.

IF INDOORS – Stay inside!
- **DUCK**, **COVER** and **HOLD** until the shaking stops!
- Avoid danger zones like glass, windows, heavy things that can fall over or down on you.

IF OUTDOORS - Stay outside!  Try to move away from buildings, power lines and street lights.

IF IN A CROWDED PUBLIC PLACE - Don't run for the door… a lot of other people will try to do that!
- Find a SAFE SPOT and avoid DANGER ZONES.
- Move away from display shelves containing objects that may fall.

IF IN A HIGH-RISE BUILDING – Stay on the same floor!
- Find a SAFE SPOT (under a desk or table).
- Move away from outside walls and windows.
- Stay in building on same floor - you may not have to evacuate.
- Realize the electricity may go out and alarms and sprinkler systems may go on.
- DO NOT use the elevators!

**IF IN A MOVING VEHICLE** - Stop as quickly and safely as you can!

- Stay in the vehicle.
- Try not to stop near or under buildings, trees, overpasses, or power lines.
- Watch for road and bridge damage and be ready for aftershocks once you drive again.

If you are trapped in an area:

- **light** - use a flashlight (if you have one) – don't use matches or lighters in case of gas leaks
- **be still** - try to stay still so you won't kick up dust
- **breathing** - cover your mouth with a piece of clothing
- **make noise** - tap on a pipe or wall so rescuers can hear you (shout only as a last resort since you could inhale a lot of dust)

## AFTER AN EARTHQUAKE:

Aftershocks - Usually not as strong but can cause more damage to weakened structures. Can occur a few more times or go on for days, months or years!

Injuries - Check yourself and people around you for injuries - do not try to move seriously injured people unless they are in danger. If you must move a person who is passed out keep their head and neck still and call for help! *(See Section 3 – TIPS ON BASIC FIRST AID)*

Light - Never use candles, matches or lighters since there might be gas leaks. Use flashlights or battery powered lanterns.

Check home - Look for structural damage -- call a professional if necessary.

Check chimney - First check from a distance to see if chimney looks normal and have a professional check it if it looks strange!

Clean up - Any flammable liquids (bleaches, gasoline, etc.) should be cleaned up immediately.

Inspect - Check all utility lines and appliances for damage:

- **smell gas or hear hissing** - open a window and leave quickly. Shut off main valve outside, if possible, and call a professional to turn back on when it's safe
- **electrical damage** - switch off power at main fuse box or circuit breaker
- **water pipes** - shut off water supply at the main valve
- **toilets** - do not use until you know sewage lines are okay

<u>Water</u> - If water is cut off or contaminated then use water from your **Disaster Supplies Kit** or other water sources.

<u>Phones</u> - Keep calls to a minimum to report emergencies since most lines will be down.

<u>Listen</u> - Keep up on news reports for the latest information.

<u>Things to avoid:</u>
- **going out** - try to stay off the roads to reduce risk
- **watch out** - watch for fallen objects and bridge or road damage
- **stay away** - unless emergency crew, police or firemen ask for your help stay away from damaged areas
- **downed power wires**

<u>Tsunami</u> - If you live near the coast, a tsunami can crash into the shorelines so listen for warnings by local authorities. *(see section on TSUNAMIS)*

<u>RED or GREEN sign in window</u> – After a disaster, Volunteers and Emergency Service personnel will be going door-to-door to check on people. By placing a sign in your window that faces the street near the door, you can let them know if you need them to STOP HERE or MOVE ON.

Either use a piece of RED or GREEN construction paper or draw a <u>big</u> RED or GREEN "X" (using a crayon or marker) on a piece of paper and tape it in the window.
- RED means STOP HERE!
- GREEN means EVERYTHING IS OKAY…MOVE ON!
- Nothing in the window would also mean STOP HERE!

<u>Recovery tips</u> - Review TIPS ON RECOVERING FROM A DISASTER at end of this Section.

# What are <u>YOU</u> gonna do about…
## AN EVACUATION?

Evacuations are pretty common and happen for a number of reasons – fires, floods, hurricanes, or chemical spills on the roads or railways.

When community evacuations become necessary, local officials provide information to the public usually through the media. Government agencies, the Red Cross and other disaster relief organizations provide emergency shelter and supplies. But, as we have said before, you should have enough food, water, clothing and emergency supplies for at least 3 days - or longer in a catastrophic disaster - in case you cannot be reached by relief efforts.

The amount of time to evacuate obviously depends on the type of disaster. Hurricanes can be tracked and allow a day or two notice to get ready, but many types of disasters happen without much notice… so prepare NOW!!

## BEFORE AN EVACUATION:

Ask & learn - Ask local emergency management officials about community evacuation plans and learn the routes that should be used.

Make a plan - Review Section 1 and develop a **Family Emergency Plan** (so you know where to meet if separated, know what schools or day cares do with kids, have a **Disaster Supplies Kit** ready to go, etc.)

Fill 'er up - Keep car fueled up if evacuation seems likely since gas stations may close during emergencies.

Learn to shut off - Know where and how to shut off electricity, gas and water at main switches and valves -- ask local utilities for instructions (and keep a wrench handy).

Review tips on basic needs - Please review TIPS ON SHELTER LIVING, TIPS ON USING HOUSEHOLD FOODS, TIPS ON WATER PURIFICA-TION and TIPS ON SANITATION OF HUMAN WASTE near end of this section to prepare yourself and family for what to expect.

## DURING AN EVACUATION:

Listen - Keep up on news reports for the latest information.

Grab & Go - Grab your **Disaster Supplies Kit** (has water, food, clothing, emergency supplies, insurance and financial records, etc. ready to go).

<u>What do I wear?</u> - Put on protective clothing (long sleeve shirt and pants) and sturdy shoes - may even want to grab a hat or cap.

<u>Shut off utilities</u> - Turn off main water valve and electricity (if authorities tell you to do so).

<u>Secure home</u> - Close and lock doors and windows, unplug appliances, protect water pipes (if freezing weather), tie down boats, etc. *(See specific types of disaster for additional tips on securing home)*

<u>Alert family / friends</u> - Let others know where you are going (or at least leave a message or note in clear view explaining where you can be found).

<u>Things to avoid</u>:
- **bad weather** - leave early enough so you are not trapped
- **shortcuts** - may be blocked -- stick to the recommended Evacuation routes
- **flooded areas** - roadways and bridges may be washed-out
- **downed power lines**

<u>Review tips on basic needs</u> - Make sure you review tips on SHELTER LIVING, USING HOUSEHOLD FOODS, WATER PURIFICATION and SANITATION OF HUMAN WASTE at end of this section to prepare your family for the unexpected.

# What are <u>YOU</u> gonna do about...
## EXTREME HEAT?

**What is Extreme Heat?** Temperatures that hover 10 degrees or more above the average high temperature for that area and last for several weeks are considered "extreme heat" or a **heat wave**. Humid and muggy conditions can make these high temperatures even more unbearable. Really dry and hot conditions can cause dust storms and low visibility. **Droughts** occur when a long period passes without enough rainfall. A heat wave combined with a drought is a very dangerous situation!

Doing too much on a hot day, spending too much time in the sun or staying too long in an overheated place can cause **heat-related illnesses**. Know the symptoms of heat illnesses and be ready to give first aid treatment. *(see HEAT-RELATED ILLNESSES in Section 3)*

## BEFORE EXTREME HEAT HITS:

Keep it cool - Tips to keep hot air out and cool air inside include...
- Close any floor heat vents nearby.
- Insulate gaps around window unit (use foam, duct tape, etc.)
- Use a circulating or box fan to spread the cool air around.
- Use aluminum foil covered cardboard in windows to reflect heat back outside.
- Use weather-stripping on doors and windowsills.
- Keep storm windows up all year to help keep cool in.

## DURING EXTREME HEAT:

Protect windows - If you hang shades, drapes, sheets, or awnings on windows you can reduce heat from entering home by as much as 80%.

Conserve power - During heat waves there are usually power shortages since everyone is trying to cool off, so stay indoors as much as possible.

Conserve water - Tips to lower water usage, esp. during drought conditions
- Check plumbing for leaks.
- Replace toilet and shower head with "low flow" versions.
- Don't leave water running while shaving, brushing teeth, washing dishes, cleaning fruit or veggies, etc.
- If washing a load of dishes or clothes, make sure it's a full load.

- Take short showers rather than filling up a bathtub.
- Limit watering lawn or washing cars -- wastes precious water.

<u>No A/C..?</u> - If you have no air conditioning, try to stay on the lowest floor out of the sunshine and use electric fans to help keep yourself cool.

<u>Eat light</u> - Light meals are best, especially fresh fruits and veggies.

<u>Drink WATER</u> - Increase your daily intake of water, esp. in dry climates (deserts and high elevations) -- you don't realize how dehydrated you get.

<u>Limit booze</u> - Even though beer and alcoholic beverages may be refreshing on a hot day, they actually cause your body to dehydrate more!

<u>What to wear</u> - Light-colored (to reflect heat) loose-fitting clothes are best... and cover as much skin as possible. Dark colors absorb the sun's heat. Also, wear a wide-brimmed hat to protect face and neck.

<u>Use sunscreen</u> - Apply lotion or cream at least 20 minutes before going out-side so skin can absorb and protect, esp. face and neck (SPF 15-30 is best but an SPF 8 should be the lowest you go). You usually burn within the first 10 minutes outside, so take care of your skin... especially <u>children</u>!! A sunburn slows the body's ability to cool itself and can be extremely dangerous.

<u>Working outdoors</u> - If you have to do yard work or other outdoor work, try to do it in the early morning hours to limit exposure in the sun. The most powerful sun is between 10 a.m. and 3 p.m. (when you burn the quickest) so limit outdoor activity during the heat of the day, if possible.

<u>Ozone alerts</u> - These can cause *serious* danger to people with breathing and respiratory problems (especially children and the elderly) so limit your time outdoors when alerts are announced on the radio, newspapers or TV.

- **ozone** - a colorless gas that is in the air we breathe and is a major element of urban smog.
- **ground-level ozone** - considered an air pollutant and can lower resistance to colds, cause problems for people with heart & lung disease, and cause coughing or throat irritation
- **ozone levels** - (also called Air Quality Index) between 0-50 are fine, but anything above 100 is extremely dangerous! When the weather is hot and sunny with little or no wind it can reach unhealthy levels.

# What are <u>YOU</u> gonna do about...
## Fires & Wildfires?

Since fire spreads so quickly, there is <u>NO</u> time to grab valuables or make a phone call!  In just <u>two</u> minutes a fire can become life threatening!  In <u>five</u> minutes a house can be engulfed in flames!

A fire's heat and smoke are more dangerous than the actual flames since you can burn your lungs by inhaling the super-hot air.  Fire produces a poisonous gas that makes you drowsy and disoriented (confused).  Instead of being awakened by a fire, you could fall into a deeper sleep!

We are going to cover two subjects here -- **FIRES** and **WILDFIRES**.  First we will discuss FIRES like you might encounter in your home or apartment, then we will cover WILDFIRES since there are many things people need to think about when living near wilderness areas.

## Before A Fire (Fire Safety Tips):

<u>INSTALL SMOKE DETECTORS!</u>  If you already have smoke detectors, clean and check them once a month and replace batteries once a year!

<u>Make a plan</u> - Review Section 1 and create an Escape Plan that includes two escape routes from every room in the house and walk through the routes with your entire family.  Also...

- Make sure windows are not nailed or painted shut.
- Make sure security bars on windows have a fire safety opening feature so they can be easily opened from the inside… and teach everyone how to open them!
- Teach everyone how to stay LOW to the floor (where air is safer) when escaping fire.
- Pick a spot outside to meet after escaping fire (meeting place).

<u>Clean up</u> - Keep storage areas clean - don't let newspapers & trash stack up.

<u>Check power sources</u> - Check electrical wiring and extension cords -- don't overload cords or outlets.  Make sure there are no exposed wires anywhere and make sure wiring doesn't touch home insulation.

<u>Use caution</u> - Never use gasoline or similar liquids indoors and never smoke around flammable liquids!

<u>Check heat sources</u> - Check furnaces, stoves, cracked or rusty furnace parts, and chimneys. Always be careful with space heaters and keep them at least 3 feet (1 m) away from flammable materials.

Know how to shut off power - Know where the circuit breaker box and gas valve is and how to turn them off, if necessary. (And always have a gas company rep turn on a main gas line.)

Install & learn A-B-C - Install A-B-C fire extinguishers in the home and teach family members how to use them. (A-B-C works on all types of fires and recommended for home - read label.)

Call local fire - Ask local fire department if they will inspect your home for fire safety and prevention.

Teach kids - Explain to children that matches and lighters are TOOLS, not toys! And teach children if they see someone playing with fire they should tell an adult right away! And finally, teach children how to report a fire and when to call 9-1-1.

Prevent common fires - Pay attention when cooking & don't smoke in bed!

## DURING A FIRE:

### If only a small fire that's not spreading too fast ...

Try to put out...? - Use a fire extinguisher or water (unless it's an electrical or grease fire) ... and never try to put out a fire that's getting out of control!

- **electrical fire** - never use water... use a fire extinguisher approved for electrical fires
- **oil or grease fire in kitchen** - smother fire with baking soda or salt (or, if burning in pan or skillet, carefully put a lid over it -- but don't try to carry pan outside!)

### If fire is spreading ...

GET OUT - DO NOT take time to try to grab anything except your family members! Once outside, do NOT try to go back in (even for pets) - let the firemen do it! Ask a neighbor to call fire department if not already called.

GET DOWN - Stay low to the ground under the smoke by crawling on your hands and knees or squat down and walk like a duck... but keep moving to find a way out!

Closed door - Using the back of your hand (not your palm) always feel the top of the door, doorknob, and the crack between the door and door frame before you open a closed door!

- **if door is cool** - leave quickly, close door behind you and crawl to an exit
- **if door is hot** – DO NOT open it ... try to find another way out

No way out - If you can't find a way out of the room you're trapped in (door is hot and too high to jump) then hang a white or light-colored sheet, towel or shirt outside a window to alert firemen.

Use stairs - Never take the elevator… always use stairs!

If YOU are on fire - If your clothes ever catch fire, **STOP** what you're doing, **DROP** to the ground, cover your face and **ROLL** until the fire goes out. Running only makes the fire burn faster!

## AFTER A FIRE:

Don't go in there - Never enter a fire-damaged building until the authorities say it is okay.

Look - Watch for signs of smoke or heat in case the fire isn't totally out.

Utilities - Have an electrician check your household wiring before you turn the power back on and DO NOT try to reconnect any utilities yourself!

Damage - Look for structural damage (roof, walls, floors, etc.) since they may be weak.

Call for help - Your local disaster relief service (Red Cross, Salvation Army, etc.) can help provide shelter, food, or personal items that were destroyed.

Insurance - Call your insurance agent or representative and…
- Keep receipts of all clean-up and repair costs (for both insurance and income taxes).
- Do not throw away any damaged goods until an official inventory has been taken by your insurance company.

If you rent - Contact your landlord since it is the owner's responsibility to prevent further loss or damage to the site.

Move your stuff - Secure your personal belongings or move them to another location, if possible.

Recovery tips - Review TIPS ON RECOVERING FROM A DISASTER at end of this Section.

*To learn more about fire safety and fire prevention visit the U.S. Fire Administration's web site www.usfa.fema.gov or contact your local fire department, state or provincial emergency management official, or your insurance agent or representative.*

**Wildfires** are intense fires that are usually caused by careless humans or lightning. Campfires, children playing with matches or lighters, and cigarettes are the most common things that cause brush fires or wildfires so please be careful when you're out in deserts, mountains, or any other heavy vegetation areas. And please don't toss cigarettes out when driving!

NEVER leave a campfire burning - make sure it is completely out using plenty of water before leaving the area. Stir the coals around with a stick or log while pouring water over them to ensure all the coals get wet and they are no longer hot. Any hot coals left unattended can be easily ignited by wind since they can stay hot for 24 - 48 hours!

When building a campfire, always choose a level site, clear away any branches and twigs several feet from the fire, and never build a fire beneath tree branches or on surface roots. Also, build at least 10 feet (3 m) from any large rocks that could be blackened by smoke or cracked from the fire's heat.

See your local Forest Service office or Ranger Station for more information on campfires and permits. (Or visit www.fs.fed.us or www.forest.ca )

## BEFORE A WILDFIRE (FIRE SAFETY TIPS):

Prepare - See WILDFIRE MITIGATION at beginning of this Section.

Learn fire laws - Ask fire authorities or the forestry office for information on fire laws (like techniques, safest times to burn in your area, etc.)

Could they find & reach you? - Make sure that fire vehicles can get to your property and that your address is clearly marked.

Safety zone - Create a 30-100 foot (9-30 m) safety zone around your home. *(see WILDFIRE MITIGATION)*

Teach kids - Explain to children that matches and lighters are TOOLS, not toys... and if they see someone playing with fire tell an adult right away! And teach kids how to report a fire and when to call 9-1-1.

Tell authorities - Report any hazardous conditions that could cause a wildfire.

Be ready to evacuate - Listen to local authorities and leave if you are told to evacuate. *(see EVACUATION)*

## DURING A WILDFIRE:

Listen - Have a radio to keep up on news, weather and evacuation routes.

Evacuate? – If you are told to leave - do so ... and IF you have time also…

- Secure your home - close windows, vents, doors, etc.
- Turn off utilities and tanks at main switches or valves, if instructed to do so.
- Turn on a light in each room to increase the visibility of your home in heavy smoke.
- Review WILDFIRE MITIGATION at front of this section.

Head downhill – Fire climbs uphill 16 times faster than on level terrain (since heat rises) so always head down when evacuating the area.

Food & water - If you prepared ahead, you'll have your **Disaster Supplies Kit** handy to **GRAB & GO**… if not, gather up enough food and water for each family member for at least 3 days or longer!

Be understanding - Please realize the firefighters main objective is getting wildfires under control and they may not be able to save every home. Try to understand and respect the firefighters' and local officials' decisions.

## AFTER A WILDFIRE:

Don't go in there - Never enter fire-damaged areas until authorities say okay.

Look - Watch for signs of smoke or heat in case the fire isn't totally out.

Utilities - Have an electrician check your household wiring before you turn the power back on and DO NOT try to reconnect any utilities yourself!

Damage - Look for structural damage (roof, walls, floors) -- may be weak.

Call for help - Your local disaster relief service (Red Cross, Salvation Army, etc.) can help provide shelter, food, or personal items that were destroyed.

Insurance - Call your insurance agent or representative and…

- Keep receipts of all clean-up and repair costs (for both insurance and income taxes).
- Do not throw away any damaged goods until an official inventory has been taken by your insurance company.

If you rent - Contact your landlord since it is the owner's responsibility to prevent further loss or damage to the site.

Move your stuff - Secure your belongings or move them to another location.

Recovery tips - See TIPS ON RECOVERING FROM A DISASTER at end of this Section.

# What are <u>YOU</u> gonna do about...
## A FLOOD?

Floods are the most common natural disaster.  Some floods develop over a period of several days, but a flash flood can cause raging waters in just a few minutes!  Mudflows are another danger triggered by flooding that can bury villages without warning (especially in mountainous regions).

Everyone is at risk from floods and flash floods, even in areas that seem harmless in dry weather.  Always listen to the radio or TV to hear the latest updates.  Some other types of radios are the NOAA (National Oceanic and Atmospheric Administration) Weather Radio and Environment Canada's Weatheradio with battery backup and a tone-alert feature that automatically alert you when a Watch or Warning has been issued.

## BEFORE A FLOOD (OR HEAVY RAIN):

<u>Prepare</u> - Review FLOOD MITIGATION at beginning of this Section.

<u>Learn the buzzwords</u> - Learn the terms / words used with floods...
- **Flood watch** - flooding is possible
- **Flash flood watch** - flash flooding is possible so move to higher ground if in a low-lying area
- **Flood warning** - flooding is occurring or will occur soon so listen to radio or TV for updates or evacuation alerts
- **Flash flood warning** - flash flood is occurring so seek higher ground on foot immediately
- **Urban and Small Stream Advisory** - flooding of small streams, streets and low-lying areas is occurring

<u>Learn risks</u> - Ask your local emergency management office if your property is a "flood-prone" or high-risk area and what you can do to mitigate (reduce risks to) your property and home.  Find out what official flood warning signals are and what to do when you hear them.  Also ask if there are dams in your area and if they could be a hazard.

<u>Be ready to evacuate</u> - Listen to local authorities and leave if you are told to evacuate. *(see EVACUATION)*

<u>Make a plan</u> - Review Section 1 to develop a **Family Emergency Plan** and **Disaster Supplies Kit**.

<u>Learn to shut off</u> - Know where and how to shut off electricity, gas and water at main switches and valves -- and ask local utilities for instructions.

<u>Get insurance...?</u> - Talk to your agent and find out more about the **National Flood Insurance Program**. *(see FLOOD MITIGATION)*

**Did you know...**

> ... you can buy federal flood insurance through most major private insurance companies and licensed property insurance agents?!
>
> ... you do <u>not</u> have to own a home to have flood insurance as long as your community participates in the **NFIP**?!
>
> ... the **NFIP** offers coverage even if you live in a flood-prone area?!
>
> ... the **NFIP** offers basement and below ground level coverage?!

<u>Put it on film</u> - Either videotape or take pictures of your home and personal belongings and store them in a safe place (like a fireproof box or a safety deposit box) along with important papers.

## DURING A FLOOD (OR HEAVY RAIN):

<u>Be aware</u> - Listen to local news and watch for flash floods especially if near streams, drainage channels, and areas known to flood.

<u>Get to higher ground</u> - If in a low-lying area, move to higher ground.

<u>Prepare to evacuate</u> – *(see EVACUATION)*, and IF time also...
- Secure your home and move important items to upper floors.
- Turn off utilities at main switches or valves if instructed by authorities and DO NOT touch electrical equipment if you are wet or standing in water!
- Fill up your car with fuel.

<u>Obey warnings</u> - If road signs, barricades, or cones are placed in areas - OBEY THEM! Most areas have fines for people who ignore these posted warnings, especially if they get stuck or flooded! DO NOT drive around barricades... find another way to get where you are going!

<u>Things to avoid</u>:
- **moving water** - 6 inches (15 cm) of moving water can knock you off your feet and 2 ft (0.6 m) of moving water can float a car
- **flooding car** - if flood waters rise around your car, get out and move to higher ground if you can do it safely! (Don't try to walk through moving water!)
- **bad weather** - leave early enough so you are not trapped
- **flooded areas** - roadways and bridges may be washed-out
- **downed power lines** - extremely dangerous in floods!!

# AFTER A FLOOD (OR HEAVY RAIN):

Things to avoid:

- **flood waters** - stay away from flood waters since may be contaminated by oil, gasoline or raw sewage or may be electrically charged from underground or downed power lines - wait for local authorities to approve returning to flooded areas
- **moving water** - 6 inches (15 cm) of moving water can knock you off your feet and 2 ft (0.6 m) of moving water can float a car
- **flooded areas** - roadways and bridges may be washed-out or weakened
- **downed power lines** - extremely dangerous and report them to the power company

Obey warnings - If road signs, barricades, or cones are placed in areas - OBEY THEM!  Most areas have fines for people who ignore these posted warnings, especially if they get stuck or flooded!  DO NOT drive around these barricades… find another way to get where you are going!

Strange critters - Watch out for snakes and other wildlife in areas that were flooded.  Don't try to care for a wounded critter since it may try to attack you... call your local animal control office or animal shelter.

Flooded food - Throw away food that has come into contact with flood waters since eating it can make you sick.

Drinking water - Wait for officials to advise when water is safe to drink.

Wash your hands - Wash hands often with <u>clean</u> water and soap since flood waters are dirty and full of germs!

Use bleach – The best thing to use for cleaning up flooded areas is household bleach since it will help kill germs.

Listen - Continue listening to your battery-powered radio for updates on weather and tips on getting assistance for housing, clothing, food, etc.

Insurance - Call your insurance agent to see if you're covered for flooding.

Mold - Consider asking a restoration professional to inspect your house for mold. *(see AIR QUALITY MITIGATION)*

Good site on cleaning basement - If you can access the Internet, visit the Seattle & King County Public Health site for tips on "Cleaning a basement after a flood" at www.metrokc.gov/health/disaster/basementflood.htm

Recovery tips - See TIPS ON RECOVERING FROM A DISASTER.

# What are <u>YOU</u> gonna do about…
## HAILSTORMS?

Hail is the largest form of precipitation that begins as tiny ice pellets and grows by colliding with supercooled water droplets as it gets tossed around violently in strong updraft winds. As the pellet continues to be tossed, it builds layer by layer until it becomes so heavy that it drops out of the sky as hailstones.

Hailstone diameters can range from 1/16 of an inch to 5 inches (2 mm to 13 mm) - basically meaning they can range in size from tiny pebbles to golfballs to grapefruits or softballs!  One of the largest hailstones ever recorded in the U.S. weighed 1.67 pounds and had a 17.5 inch (44 cm) circumference.

Hail is usually present in powerful storms like tornadoes, thunderstorms and even some winter storms mainly due to the strong winds and rapidly rising air masses needed to form hailstones.

Hail occurs across Canada but more frequently happens in the Canadian Prairies (particularly the Calgary-Medicine Hat area).  This region can expect up to 10 hailstorms a year and most of the damaging hailstorms generally occur from May to October.  The U.S. averages about 3,000 hailstorms each year across the country and a majority of the storms occur between March and June.

The worst hailstorm in Canadian history hit Calgary, Alberta in September 1991. The 30-minute downpour caused almost $400 million in insurance claims devastating crops, property and livestock.  In 1996 Alberta started a hail suppression program using aircraft that fly over developing storms and seed clouds with silver iodide particles to reduce the size of the hailstones.

## BEFORE A HAILSTORM:

*Since hailstorms are pretty localized events, it is difficult to prepare for "hail", however please review the other topics that create hailstorms (Thunderstorms, Tornadoes and Winter storms) to learn what to do and how to protect yourselves during these events!*

<u>Listen</u> - Keep up on local radio or TV weather forecasts and updates.

<u>Park it</u> - If possible, secure vehicles in a garage or under substantial cover.

<u>Bring 'em in</u> - Put pets and livestock in some type of shelter for their safety.

<u>Stay put</u> - Stay inside until the entire storm system passes.

---

# DURING A HAILSTORM:

<u>Listen</u> - Keep radio or TV tuned in for more information and updates on weather conditions and other types of warnings.

**IF INDOORS** – Stay inside until the storm passes and don't try to go out and protect your property!

**IF OUTDOORS** - Take shelter under the strongest structure you can find (especially if hailstones are large!)

**IF IN A VEHICLE** - Carefully pull over to the shoulder and seek shelter under an overpass or the closest substantial structure available.

# AFTER A HAILSTORM:

<u>Listen</u> - Continue listening to radio or TV for updates on weather.

<u>Check it out</u> - Check for damage to trees and shrubs because, if damaged, your roof most likely is too.  Also check your vehicles and structures for damage but don't put yourself in danger if storms are still active!

<u>Stop leaks</u> - Cover up holes in your roof and broken windows in your car and home to keep water out.

<u>Insurance</u> - Call your insurance agent or representative to set up a visit to your home or to take your vehicle down for inspection.

<u>Recovery tips</u> - Review TIPS ON RECOVERING FROM A DISASTER at end of this Section.

# What are <u>YOU</u> gonna do about...
## HAZARDOUS MATERIALS?

Chemical plants are one source of hazardous materials, but there are many others that exist in large industry, small businesses, and homes. There are about 500,000 products that could pose a physical or health hazard -- things ranging from waste produced by a petroleum refinery to materials used by the dry cleaners to pesticides stored in your home.

Most hazardous materials are transported around the country by road, rail and through pipelines potentially causing spills on highways, near railroad tracks or underground. Many U.S. communities have a **Local Emergency Planning Committee (LEPC)** that keeps local planners, companies and members of the community informed of potential risks. All companies that have hazardous chemicals must report to the LEPC every year and the public is encouraged to get involved. We [the public] should all learn more about hazardous materials and how they can affect our lives so contact your emergency management office to learn more.

We're going to cover two topics here -- **HAZARDOUS MATERIALS DISASTER** (where a spill or incident affects an area or community) and **HOUSEHOLD CHEMICAL EMERGENCIES** (how to handle products and react if there's an emergency in the home). Also, please review the TERRORISM topic since it covers several biological and chemical agents that are also classed as "hazardous materials".

## BEFORE A HAZARDOUS MATERIALS DISASTER:

<u>Learn the buzzwords</u> - Ask your local officials about emergency warning procedures and terms...

- **Outdoor warning sirens or horns** - ask what they mean and what to listen for
- **Emergency Alert System (EAS)** - information and alerts via TV and radio
- **"All-call" telephoning** - an automated system for sending recorded messages via telephone
- **Residential route alerting** - messages announced from vehicles equipped with public address systems (loud speakers on top of car or van)

<u>Learn risks</u> - Ask your Local Emergency Planning Committee (LEPC), Emergency Management Office, or Fire Department about community plans for responding to hazardous materials accident at a plant or a transportation accident involving hazardous materials. Also ask where large quantities of extremely hazardous substances are stored and where they are used.

Make a plan - Use LEPC's or agency's information to see if your family is at risk, especially people living close to freeways, railroads, or factories which produce or transport toxic waste. And review Section 1 to develop a **Family Emergency Plan** and **Disaster Supplies Kit**.

Take a tour - Arrange a neighborhood tour of industries that produce or transport toxic waste and include neighbors, local officials and the media.

Pick a room - It could take authorities time to determine what (if any) the hazardous material is so pick a room in advance that your family could use if you are told to stay indoors for several hours. It's best to pick an internal room where you could block out air, if instructed to do so. To save critical time consider measuring and cutting plastic sheets in advance for each opening (vents, windows, and doors). Remember, a toilet is usually vented meaning outside air comes in constantly or when flushed (depends on design) - just FYI in case you choose bathroom as a safe room.

Calculate air for room - Keep in mind people can stay in a sealed off room for only so long (or you'll run out of air!) FEMA suggests 10 square feet of floor space per person (like 5ft x 2ft / 1.5m x 0.6m ) will provide enough air to prevent carbon dioxide buildup for up to 5 hours.

Be ready to evacuate - Listen to local authorities and leave if you are told to evacuate. *(see EVACUATION)*

## DURING A HAZARDOUS MATERIALS DISASTER:

Call for help - If you see a hazardous materials accident, call 9-1-1, local emergency number, or the fire department.

Listen - Keep radio or TV tuned in for more information, especially if you hear a warning signal... and stay calm!

IF INDOORS – If told to stay inside, do it - and ...
- Close your windows, vents, and fireplace dampers and turn off A/C or heat and fans to reduce air drawn in from outside.
- Keep a radio with you at all times.
- Grab your **Disaster Supplies Kit** and get to a closed off room.
- Seal gaps under doorways and windows with wet towels or plastic and duct tape (see above tips on picking a room and calculating air!)

IF OUTDOORS - Stay upstream, uphill, or upwind from the disaster since hazardous materials can be carried by wind and water quickly. Try to get at least ½ mile or kilometer away or as far away as possible!

**IF IN A VEHICLE** - Close your windows and shut off vents to reduce risk.

Stay away - Get away from the accident site to avoid contamination.

Evacuate...? - If you are told to evacuate… do it!  If officials say you have time, close windows, shut vents and turn off attic fans.  *(see EVACUATION)*

What to wear - Keep your body fully covered and wear gloves, socks and shoes. (Even though these may not keep you totally safe, it can help!)

Things to avoid:
- **chemicals** - spilled liquid materials or airborne mists
- **contaminated food or water** - don't eat or drink any food or water that may have been exposed to hazardous materials

## AFTER A HAZARDOUS MATERIALS DISASTER:

Don't go there - Do not return home until local authorities say it is safe.

Air out - Open windows, vents and turn on fans in your home.

Listen - Keep up with local reports from either the radio or TV.

Clean up - A person, critter or item that has been exposed to a hazardous chemical could spread it.
- **decontamination** - follow instructions from local authorities since it depends on the chemical.  You may need to shower or rinse off or may be told to stay away from water - check first!
- **strange symptoms** - if unusual symptoms show up, get to a hospital or medical expert right away!  Remove contaminated clothing and put on fresh, loose, warm clothing and listen to local reports on the radio.
- **store clothes & shoes** - put exposed clothing and shoes in tightly sealed containers/bags without touching other materials and call local authorities to ask how to get rid of them
- **tell people you've been exposed** - tell everyone who comes in contact with you that you may have been exposed to a toxic substance
- **land and property** - ask authorities how to clean the area

Strange vapors or danger - Report any strange vapors or other dangers to the local authorities immediately.

To learn more about hazardous materials, visit the U.S. Environmental Protection Agency's Chemical Emergency Preparedness and Prevention

Office (CEPPO) at www.epa.gov *(do a Search on CEPPO or "Browse Topics" then several to choose from like Emergencies, Pollutants/Toxics, Wastes, etc.)*

Or visit Environment Canada at www.ec.gc.ca *(Click on "Topics" then several to choose from like Environmental Emergencies, Pollution, Waste Management, etc.)* ... or the Canadian Transport Emergency Centre of the Department of Transport at www.tc.gc.ca/canutec/

## BEFORE A HOUSEHOLD CHEMICAL EMERGENCY:

Learn risks - Call your Local public health department or the Environmental Protection Agency for information about hazardous household materials.

Read labels - Always read product labels for proper use and disposal of chemicals.

Recycle it? - Call your local recycling center or collection site to ask what chemicals can be recycled or dropped off for disposal -- many centers take things like car batteries, oil, tires, paint or thinners, etc.

Store it - Keep all chemicals and household cleaners in safe, secure locations out of reach of small children.

Put it out - Don't smoke while using household chemicals.

## DURING A HOUSEHOLD CHEMICAL EMERGENCY:

Call for help - **Call your local Poison Control Center, 9-1-1, fire department, hospital or emergency medical services.**

First aid tips - Follow instructions on label and see Basic First Aid tips for POISONING in Section 3.

# What are <u>YOU</u> gonna do about...
## HURRICANES, CYCLONES & TYPHOONS?

Hurricane season in North America is generally between June and November. Hurricanes are tropical cyclones with torrential rains and winds of 74 - 155 miles per hour (120 - 250 km/h) or faster. These winds blow in a counter-clockwise direction (or clockwise in the Southern Hemisphere) around a center "eye". The "eye" is usually 20 to 30 miles (32 to 48 km) wide, and the storm may be spread out as far as 400 miles (640 km)!

As the hurricane approaches the coast, a huge dome of water (called a storm surge) will crash into the coastline. Nine out of ten people killed in hurricanes are victims of storm surge. Hurricanes can also cause tornadoes, heavy rains and flooding.

### What's with all the different names?

You may have heard different words used to describe different storms depending on where you live in the world. It is confusing but hopefully we can help explain all the different names... and hopefully we don't make any weather specialists angry.

<u>Cyclone</u> - an atmospheric disturbance with masses of air rapidly rotating around a low-pressure center... (sort of like a dust devil or a tornado)

<u>Tropical Depression</u> - maximum surface winds of less than 39 miles per hour (62 km/h) over tropical or sub-tropical waters with storms and circular winds

<u>Tropical Storm</u> - the tropical cyclone is labeled a Tropical Storm if winds are between 39-73 mph (62 - 117 km/h) and given a name to track it

<u>Hurricane, Typhoon, Tropical cyclone</u> - surface winds are higher than 74 mph (120 km/h)... and depending on where it is happening will determine what it is called

### Where in the world do they use these names?

*(Please note: We are only listing a <u>few</u> major countries or areas for each!)*

<u>Cyclone</u> - used in several parts of the world - **Indian Ocean, Australia, Africa, SW and southern Pacific Ocean**

<u>Hurricane</u> - used in North Atlantic Ocean, Northeast Pacific Ocean (east of the dateline), or South Pacific Ocean (east of 160) - **both coasts of North America, Puerto Rico, Caribbean Islands, and Central America**

<u>Typhoon</u> - used in Northwest Pacific Ocean west of the dateline - **Guam, Marshall Islands, Japan, Philippines, Hong Kong, coastal Asia**

<u>Tropical cyclone</u> - used in Southwest Pacific Ocean west of 160E or most of the Indian Ocean - **Australia, Indonesia, Madagascar, Africa, Middle East**

Hurricanes are classed into five categories based on wind speeds, central pressure, and damage potential. The chart below is called the Saffir-Simpson Hurricane Scale with some examples of damage provided by FEMA:

| Scale # (Category) | Sustained Winds | Damage | Storm Surge |
|---|---|---|---|
| 1 | 74-95 mph 119-153 km/h | **Minimal**: Untied mobile homes, vegetation & signs | 4-5 ft 1.2-1.5 m |
| 2 | 96-110 mph 154-177 km/h | **Moderate**: All mobile homes, roofs, small crafts, flooding | 6-8 ft 1.8-2.4 m |
| 3 | 111-130 mph 178-209 km/h | **Extensive**: Small buildings, low-lying roads cut off | 9-12 ft 2.7-3.6 m |
| 4 | 131-155 mph 210-249 km/h | **Extreme**: Roofs and mobile homes destroyed, trees down, beach homes flooded | 13-18 ft 3.9-5.4 m |
| 5 | > 155 mph > 250 km/h | **Catastrophic**: Most bldgs and vegetation destroyed, major roads cut off, homes flooded | > 18 ft > 5.4 m |

## BEFORE A HURRICANE:

<u>Prepare</u> - Review WIND, FLOOD, and LIGHTNING MITIGATION at beginning of this Section.

<u>Learn the buzzwords</u> - Learn the terms / words used with hurricanes...
- **Hurricane/Tropical Storm Watch** - hurricane/tropical storm is possible within 36 hours so listen to TV and radio updates
- **Hurricane/Tropical Storm Warning** - hurricane/tropical storm is expected within 24 hours -- may be told to evacuate (if so, do it) and listen to radio or TV for updates
- **Short term Watches and Warnings** - warnings provide detailed information on specific hurricane threats (like flash floods and tornadoes)

<u>Listen</u> - Keep local radio or TV tuned in for weather forecasts and updates. (Some other radios to consider are Environment Canada's Weatheradio and NOAA's Weather Radio with battery backup and tone-alert feature that automatically alert you when a Watch or Warning has been issued.)

<u>Be ready to evacuate</u> - Listen to local authorities and leave if you are told to evacuate. *(see EVACUATION)*

Make a plan - Review Section 1 to develop a **Family Emergency Plan** and **Disaster Supplies Kit**.

Learn to shut off - Know where and how to shut off electricity, gas and water at main switches and valves -- ask local utilities for instructions.

Batten down - Make plans to protect your property with storm shutters or board up windows with plywood that is measured to fit your windows. Tape does not prevent windows from breaking. *(See WIND MITIGATION)*

Get insurance...? - Talk to your agent and find out more about the **National Flood Insurance Program**. *(see FLOOD MITIGATION)*

Put it on film - Either videotape or take pictures of your home and personal belongings and store them in a safe place (like a fireproof box or a safety deposit box) along with important papers.

## DURING A HURRICANE THREAT:

Listen - Have a battery-operated radio available to keep up on news reports and evacuation routes.

Evacuate? – If you are told to evacuate - do it! *(see EVACUATION)* And if you have time also…
- Secure your home - close storm shutters or put up boards on windows, moor your boat, and secure outdoor objects or put them inside since winds will blow them around.
- Turn off utilities at main switches or valves, if instructed.
- Fill up your car with fuel.

Food & water - If you prepared ahead, you'll have your **Disaster Supplies Kit** handy to GRAB & GO… if not, gather up enough food and water for each family member for at least 3 days!

**IF INDOORS** – Stay inside!
- Find a SAFE SPOT - get to a small interior room, closet or hallway ... or lie on the floor under a heavy desk or table.

**IF IN A MULTI-STORY BUILDING** – Go to the first or second floor!
- Find a SAFE SPOT - get to a small interior room or hallway ... or lie on the floor under heavy desk or table.
- Move away from outside walls and windows.
- Realize the electricity may go out and alarms and sprinkler systems may go on.

**Pets** - Make arrangements for your pets since most shelters won't allow them.

Things to avoid:

- **moving water** - 6 inches (15 cm) of moving water can knock you off your feet and 2 ft (0.6 m) of moving water can float a car
- **flooding car** - if flood waters rise around your car, get out and move to higher ground if you can safely! (Don't try to walk through moving water!)
- **bad weather** - leave early enough so you are not trapped
- **flooded areas** - roadways and bridges may be washed-out
- **downed power lines** - extremely dangerous in floods!!

Stay indoors - If you do not evacuate, stay indoors and stay away from glass doors and windows. Keep curtains and blinds closed and remember, a lull in the storm could only be the middle of the storm (the "eye") and winds can start again! Keep listening to radio or TV reports.

Limit phone calls - Only use telephones in an emergency so it keeps lines open for local authorities!

## AFTER A HURRICANE:

Stay put - Stay where you are (if you're in a safe location) and don't return home (if you've been evacuated) until local authorities say it's okay.

Listen - Continue listening to your battery-powered radio for updates on weather and tips on getting assistance for housing, clothing, food, etc.

Stick together - Keep family together since this is a very stressful time and try to find chores for children so they feel they're helping with the situation.

Things to avoid:

- **flood waters** - stay away from flood waters since it may be contaminated by oil, gasoline or raw sewage or may be electrically charged from underground or downed power lines - wait for local authorities to approve returning to flooded areas
- **moving water** - 6 inches (15 cm) of moving water can knock you off your feet and 2 ft (0.6 m) of moving water can float a car
- **flooded areas** - roadways and bridges may be washed-out or weakened
- **downed power lines** - extremely dangerous and report them to power company

Drinking water - Wait for officials to advise when water is okay to drink!

RED or GREEN sign in window – After a disaster, Volunteers and Emergency Service personnel usually go door-to-door to check on people. By placing a sign in your window that faces the street near the door, you can let them know if you need them to STOP HERE or MOVE ON (if home is still standing!). Either use a piece of RED or GREEN construction paper or draw a big RED or GREEN "X" (using a crayon or marker) on a piece of paper and tape it in the window.

- RED means STOP HERE!
- GREEN means EVERYTHING IS OKAY…MOVE ON!
- Nothing in the window would also mean STOP HERE!

Flooded food - Throw away any food that has come into contact with flood waters since eating it can make you sick!

Wash your hands - Use clean water and soap when washing hands.

Use bleach – The best thing to use for cleaning up flooded areas is household bleach since it will help kill germs.

Insurance - Call your insurance agent to set up a visit to your home.

Mold - Consider asking a restoration professional to inspect your house for mold. *(see AIR QUALITY MITIGATION)*

Donations – Lots of people want to help victims of a hurricane and here are some tips…

- **wait & see** - don't donate food, clothing or other personal items unless they are specifically requested
- **money** - donations to a known disaster relief group, like the Red Cross, is always helpful
- **volunteers** - if local authorities ask for your help, bring your own water, food and sleeping gear

Recovery tips - Review TIPS ON RECOVERING FROM A DISASTER at end of this Section.

# What are __YOU__ gonna do about...
## A Nuclear Power Plant Emergency
## (or a Nuclear Incident)?

The World Nuclear Association reports as of March 2004 that 440 nuclear power reactors in 32 countries produce 16% of the total electricity generated worldwide with 134 more reactors under construction or planned. There are over 100 commercial power plants across the U.S. and 20 power stations in Canada (18 in Ontario, 1 in Quebec and 1 in New Brunswick) meaning millions of citizens live within 10 miles (16 km) of an operating reactor.

Even though national governments and associations monitor and regulate construction and operation of plants, accidents are possible and do happen. An accident could result in dangerous levels of radiation that could affect the health and safety of the public living near the nuclear power plant, as well as people up to 200 miles (320 km) away depending on winds and weather -- so millions and millions of North Americans could potentially be affected!

**Please note**: Some other types of incidents involving possible radiation exposure may be a "radiological" event (like a "dirty bomb"), a "dirty nuke" (a suitcase-sized nuclear device), or a "weapon of mass destruction" (like a nuclear missile). "Dirty bombs" are briefly covered in the next topic called TERRORISM, but please review the next few pages before moving on. The chances of a nuclear emergency happening are remote but learn the risks, make a plan so you know how to react, and listen to authorities.

### How is radiation detected?

You cannot see or smell radiation - scientists use special instruments that can detect even the smallest levels of radiation. If radiation is released, authorities from Federal and State or Provincial governments and the utility will monitor the levels of radioactivity to determine the potential danger so they can protect the public.

### What is the most dangerous part of a nuclear accident?

**Radioactive iodine** - nuclear reactors contain many different radioactive products, but the most dangerous one is radioactive iodine which, once absorbed, can damage cells of the thyroid gland. The greatest population that suffers in a nuclear accident is **children** (including underbornunborn babies) since their thyroid is so active, but all people are at risk of absorbing radioactive iodine.

### How can I be protected from radioactive iodine?

**Potassium iodide (KI)** - can be purchased over-the-counter now (usually from companies selling disaster-related kits) and is known to be an effective thyroid-blocking agent. In other words, it fills up the thyroid with good iodine that keeps the radioactive iodine from being absorbed into our bodies.

## What if I am allergic to iodine?

According to the United States Nuclear Regulatory Commission Office of Nuclear Material Safety and Safeguards, the FDA suggests that risks of allergic reaction to potassium iodide are minimal compared to subjecting yourself to cancer from radioactive iodine. Ask your doctor or pharmacist what you should keep on hand in the event of an allergic reaction.

Many European countries stockpile potassium iodide (KI), especially since the Chernobyl incident. Several states within the U.S. are considering or already have stockpiles of KI ready in case of a nuclear power plant accident or incident as part of their Emergency Planning.

*As of March 2003, the FDA has approved 3 KI products - Thyro-Block, Iosat, and Thyrosafe. To learn more visit www.fda.gov/cder/drugprepare/KI_Q&A.htm or www.bt.cdc.gov/radiation/ki.asp*

## Community Planning for Emergencies

Local, state and provincial governments, Federal agencies and utilities have developed emergency response plans in the event of a nuclear power plant accident. *(Per FEMA's Nuclear Power Plant Emergency Backgrounder)*

**U.S.** plans define 2 "emergency planning zones" (EPZs)

- **Plume Exposure EPZ** - a 10-mile radius from nuclear plant where people could possibly be harmed by radiation exposure
  *NOTE: People within 10-mile radius are given emergency inform-ation about radiation, evacuation routes, special arrangements for handicapped, etc. via brochures, phone books, and utility bills.*

- **Ingestion Exposure EPZ** - about a 50-mile radius from plant where accidentally released radioactive materials could contaminate water supplies, food crops and livestock

**Canada's** Provincial Nuclear Emergency Response Plans define 3 "zones"
*(Per Ontario Ministry of Public Safety & Security EMO PNERP Backgrounder)*

- **Contiguous Zone** - extends approximately 3 kilometres from nuclear facility where evacuation and sheltering may be ordered

- **Primary Zone** - extends approximately 10 kilometres from the nuclear facility where evacuation and sheltering may be ordered

- **Secondary Zone** - extends approximately 50 kilometres from the nuclear facility where radioactive contamination could cause monitoring and/or bans on some food and water sources
  *NOTE: Public Education brochures are available to residents and businesses within the Primary Zone (10 km) of each nuclear facility.*

# 3 Ways to Reduce Radiation Exposure

DISTANCE - The more distance between you and the source of radiation, the less radiation you will receive - that's why in a serious nuclear accident you are told to evacuate.

SHIELDING - Heavy, dense materials between you and radiation is best - this is why you want to stay indoors since the walls in your home should be good enough to protect you in some cases... but listen to radio and TV to learn if you need to evacuate!

TIME - Most radioactivity loses its strength rather quickly so by limiting your time near the source of radiation, it reduces the amount you receive.

# Before a Nuclear Emergency or Incident:

Learn the buzzwords - Know terms used in both countries to describe a nuclear emergency: **U.S. / (Canada)**...

- **Notification of Unusual Event / (Reportable Event)** - a small problem has occurred at the plant. No radiation leak is expected. Federal, state/provincial and county/municipal officials will be told right away. No action on your part will be necessary.

- **Alert / (Abnormal Incident)** - a small problem has occurred, and small amounts of radiation could leak inside the plant. This will not affect you and you shouldn't have to do anything.

- **Site Area Emergency / (Onsite Emergency)** - a more serious problem... small amounts of radiation could leak from the plant. If necessary, officials will act to ensure public safety. Area sirens may be sounded and listen to your radio or TV for information.

- **General Emergency / (General Emergency)** - the MOST serious problem... radiation could leak outside the plant and off the plant site. In most cases sirens will sound so listen to local radio or TV for reports. State/Provincial and county/municipal officials will act to assure public safety and be prepared to follow their instructions!

Learn signals - Ask about your community's warning system and pay attention to "test" dates to learn if you can HEAR it. Nuclear power plants are required to install sirens and other warning devices to cover a 10-mile area around the plant in the U.S. (If you live outside the 10-mile area you will probably learn of the event through local TV and radio, but just be aware winds and weather can impact areas as far as 200 miles [320 km] away!!)

Learn risks - Ask the power company operating the nuclear power plant for brochures and information (which they or government sends automatically to people within a 10-mile [10-km in Canada] radius of the plant).

<u>Make a plan</u> - Review Section 1 to develop a **Family Emergency Plan** and **Disaster Supplies Kit**. Double check on emergency plans for schools, day cares or places family may be and where they'll go if evacuated.

<u>Be ready to evacuate</u> - Listen to local authorities and leave if you are told to evacuate. *(see EVACUATION)*

## DURING A NUCLEAR EMERGENCY OR INCIDENT:

<u>Stay calm</u> - Not all accidents release radiation - may only be in power plant!

<u>Listen</u> - Turn on radio or TV. Authorities will give specific instructions and information... pay attention to what THEY tell you rather than what is written in this Manual since they know the facts for each specific incident.

<u>Evacuate..?</u> - Only leave if told to do so by local authorities ... and ...
- Grab your **Disaster Supplies Kit**.
- Close doors, windows and fireplace damper.
- Close car windows and vents and use "re-circulating" air.
- Keep listening to radio for evacuation routes & updates.

### As long as you are NOT told to evacuate, do the following...

**IF INDOORS** - If you are not told to evacuate, stay inside!
- Close doors and windows and your fireplace damper.
- Turn off air conditioner, ventilation fans, furnace and other intakes (they pull in air from outside).
- Go to a basement or underground area (if possible).
- Keep a battery-operated radio with you to hear updates.
- Stay inside until authorities tell you it is safe to go out!

**IF OUTDOORS** - Get indoors as soon as possible!
- Cover mouth and nose with a cloth or handkerchief.
- Once inside, remove clothing, take a good shower and put on fresh clothing and different shoes. Put clothes and shoes you were wearing in plastic bags, seal and store. Local authorities can tell you what to do with bags.

**IF IN A VEHICLE** - Keep windows up, close vents, use "recirculating" air and keep listening to radio for updates. If possible, drive away from site.

**IF AN EXPLOSION OR BLAST** - (like from a possible nuclear device)
- Do NOT look directly at flash, blast or fireball!
- Stay low and watch out for flying debris or fires.

- A blast could create an electromagnetic pulse (EMP) that may fry electronics connected to wires or antennas like cell phones, computers, cars, etc.  May harm people with pacemakers.

Food - Put food in covered containers or in the refrigerator -- any food that was not in a covered container should be washed first.

Pets & livestock - Get them indoors or in shelters with clean food and water that has not been exposed to air-borne radiation (food and water that has been stored), especially milk-producing animals.

Take potassium iodide..? - IF radioactive iodine has been released into the air from a power plant accident, some states *may* decide to provide KI pills mentioned at beginning of this topic to people in a 10-mile radius.

(In June 2002 President Bush signed a provision that gave state and local governments supplies of potassium iodide for people within 20 miles of a nuclear power plant, increasing protection beyond the Nuclear Regulatory Commission's current 10-mile radius.[4]  This is at the option of state and local government and realize it will take time for them to disperse to citizens ... unless you prepare in advance and keep KI handy for such emergencies, but only take if officials confirm radiation was released outside the plant!)

*NOTE:  Take KI pills ONLY as directed by state, provincial or local public health authorities and follow instructions on the package exactly!  (See pages 74-75.)*

## AFTER A NUCLEAR EMERGENCY OR INCIDENT:

Listen - Keep radio and TV tuned in -- stay in until authorities say all clear.

Clean up - If you were possibly exposed to radiation...
- **store clothes & shoes** - put clothing and shoes in tightly sealed containers or plastic bags and ask health officials what to do with them
- **shower** - wash body & hair to remove radioactive particles
- **land and property** - ask authorities how to clean up area

Weird symptoms - Seek medical attention if you have symptoms like upset stomach or feel queasy after a reported incident since it could be related to radiation exposure. *(see page 105 for more about radiation sickness)*

Gardens – Authorities will provide information concerning safety of farm and homegrown products - or check with agricultural extension agent.

Crops - Unharvested crops are hard to protect but crops that are already harvested should be stored inside, if possible.

Milk - Local officials should inspect milk from cows and goats before using.

# What are <u>YOU</u> gonna do about...
## TERRORISM?

Terrorism is the use of force or violence against persons or property usually for emotional or political reasons or for ransom. The main goal of terrorists is to create public fear and panic.

Obviously there is a lot of anxiety since the September 11, 2001 attacks on the U.S., however, being afraid or worrying is very unhealthy - especially about something you have little control over. But remember, terrorist attacks are a very <u>low risk</u> possibility. Let's put a few "risks" in perspective ... the chances of having high blood pressure is 1 in 4 ... the odds of dying from cancer is 1 in 500 ... and the odds of dying from anthrax is 1 in 56 <u>million</u>!

People need to remain calm about the threat of terrorist attacks and learn about some of the types, how to prepare for them, and what to expect in some cases. Discuss this with everyone - even the kids so they can talk about their feelings too. Stay current on news but don't obsess over it ... and just be aware of your surroundings as you go about your daily routines.

One type of terrorism that we can help prevent is the use of guns and bombs by children and youth against other groups of children at schools. A key solution to stopping this type of school violence is through communication, education and awareness – and it starts within <u>the FAMILY</u>!

The Federal Bureau of Investigation categorizes terrorism in two ways:

<u>Domestic terrorism</u> - terrorist activities are directed at certain groups or parts of the government within the U.S. without foreign direction.

> Some examples of domestic terrorism include shootings and bomb threats at schools, the Oklahoma City bombing of the Federal Building, and the letters mailed to various groups with a white powdery substance (anthrax scares).

<u>International terrorism</u> - terrorist activities are foreign-based by countries or groups outside the U.S.

> Some examples of international terrorism include bombings like the U.S.S. Cole in Yemen and U.S. Embassies in other countries, the attacks on the Pentagon and World Trade Center, hostage situations with civilians in various countries, or threats with weapons of mass destruction.

Until recently, most terrorist attacks involved bombs, guns, kidnappings and hijackings, but some other forms of terrorism involve <u>cyber attacks</u>, <u>biological</u> or <u>chemical</u> agents, <u>radiological</u> or <u>nuclear</u> devices (the last 4 now considered <u>weapons of mass destruction</u>).

<u>Cyber attacks</u> - computer-based attacks from individuals or terrorist groups causing severe problems for government, businesses and public in general (sometimes causing or leading to injury and death)

<u>Biological agents</u> - infectious microbes (tiny life forms), germs or other substances that occur naturally or are "designed" to produce illness or death in people, animals or plants -- can be inhaled, enter through a cut in the skin, or swallowed when eating or drinking

<u>Chemical agents</u> - poisonous vapors, liquids or solids that can kill or slow down or weaken people, destroy livestock or crops -- can be absorbed through the skin, swallowed or inhaled

<u>Radiological threat or device</u> - a "dirty bomb" or RDD uses conventional explosives to spread radioactive materials over a general or targeted area

<u>Nuclear device</u> - a bomb or missile using weapons grade uranium or plutonium (*please note, we covered nuclear-related incidents on pages 74-78*)

<u>Weapons of mass destruction</u> (WMD) - chemical, biological, radiological, and nuclear devices are now all classed as WMDs

Terrorism is quite an extensive topic now -- below we are listing some basic things to do before any type of terrorist attack. Then we'll cover specific types shown above in red - including what to do BEFORE, DURING and AFTER each and where to find more information. We also threw in some tips for handling "bomb threats" or "suspicious packages".

Keep in mind, the best thing you can do about terrorism is prepare yourself and your family for the unexpected, so please review this topic and the previous one on "nuclear" threats. By learning about potential threats, we are all better prepared to know how to react if the unthinkable happens.

## BEFORE <u>ANY</u> TYPE OF TERRORIST ATTACK:

<u>BE AWARE!</u> - You should always be aware of your surroundings and report any suspicious activities to local authorities.

<u>Stay current on alerts</u> - Canada's PSEPC *(pages 196-199)* and the U.S.'s Department of Homeland Security *(APPENDIX A)* post alerts on the Internet

<u>Learn "Threat Levels"</u> - Review the Homeland Security Advisory System *(pages 201-208)* to see what your family or business should do at each color

<u>Know the targets</u> - Terrorists usually prefer to pick targets that bring little damage to themselves and areas that are easy to access by the public (like international airports, military and government buildings, major events,

schools, etc.) Some other high risk targets include water and food supplies, utility companies (esp. nuclear power plants) and high-profile landmarks.

Things to watch out for:
- **unknown packages** - DO NOT accept a package or case from a stranger
- **unattended bags** - DO NOT leave bags or purses alone (especially when traveling) and NEVER ask strangers to watch your stuff!
- **emergency exits** - always be aware of where Emergency EXITS are… just casually look around for the signs since most are marked well in public places

Make a plan - Review Section 1 to develop a **Family Emergency Plan** and **Disaster Supplies Kit**.


## CYBER ATTACKS

There are 3 key risk factors related to information technologies (IT) systems:

- A direct attack against a system "through the wires" alone (called hacking) -- meaning an attacker or user "hacks" in or gains "**access**" to restricted data and operations.

- An attack can be a physical assault against a critical IT element -- meaning an attacker changes or destroys data, modifies programs or takes control of a system (basically can cause a loss of data "**integrity**" = data is no good).

- The attack can be from the inside -- meaning private information could get in the wrong hands and become public or identities stolen (basically "**confidentiality**" is broken = data is no longer secure or private).

Cyber attacks target computer networks that run government, financial, health, emergency medical services, public safety, telecommunications, transportation and utility systems - also known as "critical infrastructure".

Because technologies have improved our access to information, we have opened ourselves up for attacks by our enemies to destroy or alter this data. Cyberterrorism is different than computer crime or "hactivism" (which can be costly and a pain to fix but doesn't threaten lives or public safety.)

Cyberterrorism is usually done with a minimal loss of life but there are certain terrorist groups that could potentially use cyber attacks to cause human casualties or fear by disrupting transportation or public safety systems.

Again, we are not trying to cause worry or panic, but understand the possibility exists and services could be disrupted or cut off or man-made disasters could happen due to cyber attacks. For example, services like banking, gas pumps, or internet access could be down or slow. And some emergency planners are concerned a cyber attack combined with a physical act of terrorism (like a "dirty bomb" or releasing a chemical or biological agent) could potentially interfere with response capabilities.

Most countries have agencies committed to securing and monitoring "critical infrastructure" and share information with each other on a regular basis. As we mentioned earlier, the public should stay current on alerts and news relating to national security by visiting the U.S.'s Department of Homeland Security site at www.dhs.gov and Canada's Public Safety and Emergency Preparedness Canada (formerly OCIPEP) site at www.ocipep.gc.ca .


## BIOLOGICAL AGENTS

Biological agents are actually tiny life forms or germs that can occur naturally in plants, animals and soils or can be developed for scientific or military purposes. Many biological agents affect humans by being inhaled, absorbed into the skin through a cut, or by swallowing contaminated food or water. But there are things that make it difficult for some biological agents to live like sunlight (ultraviolet light) or dry conditions. Wind could carry agents long distances but also spreads it out making it less effective.

Many animals and insects carry diseases that affect humans but most don't make us sick when eaten or inhaled because our immune systems are strong enough to fight them. But, if a person's immune system is weak (like in babies or the elderly), it's possible that person could become sick or die.


### What biological agents could be used in an attack?

There are **3 basic groups** of biological agents that could be weaponized and used in an attack (but realize there are some that occur naturally too):

- **Bacteria** - tiny life forms that reproduce by simple division and are easy to grow -- the diseases they spread are killed by a strong or boosted immune system or antibiotics (if necessary)

- **Viruses** - organisms that need living cells to reproduce and are dependent on the body they infect -- most diseases caused by viruses don't respond to antibiotics but sometimes antiviral drugs work (and a boosted immune system may help fight the invading organisms but depends on the type of virus)

- **Toxins** - poisonous substances found in and extracted from living plants, animals or microorganisms; some toxins can be produced or altered -- some toxins can be treated with specific antitoxins and selected drugs

Remember, biological weapons - or germ warfare - have been around since World War I so it's not anything new ... it's unfortunate we have to discuss it at length, but try not to let this topic frighten you. Educate yourselves about the types and where to find more information so you are prepared to react.

## How could biological agents be used in an attack?

As mentioned earlier, most biological agents break down when exposed to sunlight or other conditions, and they are very hard to grow and maintain.

There are 3 ways biological agents could be spread:

- **Aerosols** - dispersed or spread into air by a number of methods forming a fine mist that could drift for miles
- **Animals and insects** - some diseases can be carried and spread by critters like birds, mice or rodents, mosquitoes, fleas, or livestock -- a process also known as "agroterrorism"
- **Food and water contamination** - most organisms and toxins are killed or deactivated when we cook food and boil or treat water but some may continue living

Some biological agents could remain in the environment and cause problems long after they are released. But keep in mind, both the Center for Disease Control and Environmental Protection Agency are working closely with various Departments of Defense and Energy and many other agencies around the country to monitor systems and security and develop plans. The same can be said for Health Canada and many Canadian government agencies.

The CDC also suggests citizens not be frightened into thinking they need a gas mask or be concerned about food and water sources. In the event of a public health emergency, local and federal health departments will tell people what actions need to taken.

## What are the names of some biological agents and what can they do?

According to the Center for Disease Control's Public Health Emergency Preparedness and Response web site, there are many types of biological diseases and agents - in fact, too many to list here. The CDC has categorized biological agents into 3 groups (A, B and C). The diseases and agents listed in Category A are considered "highest priority" and rarely seen in the United States. Since most of the same agents were also listed on Health Canada's Emergency Preparedness and Response site, we decided to cover 7 specific agents (plus "ricin" from Category B) in alphabetical order.

**Anthrax** - is an infection caused by bacteria (*Bacillus anthracis*) found naturally in soil where it can live for years. The bacteria form a protective coat around themselves called spores which are very tiny, invisible to the naked eye, and odorless. Anthrax is most common in cows and sheep but can also infect humans (primarily people who work with hoofed animals).

**How it spreads:** Anthrax cannot spread from person to person. People come into contact with bacteria by breathing in spores (**inhalation**), by getting it through a cut in skin (**cutaneous**) or by eating something containing bacteria - like undercooked meat from an infected animal (**gastrointestinal**).

**Signs & Symptoms:** Signs depend on type of anthrax you're exposed to:

- Inhalation - most serious form - first signs similar to cold or flu (sore throat, fever and extremely tired but <u>no</u> runny nose) -- after several days may lead to severe breathing problems, shock, then possibly death
- Cutaneous - least serious form - first symptom is a small painless sore that turns into a blister -- a day or two later blister forms a black scab in the center
- Gastrointestinal - at first nausea, loss of appetite, puking, and fever -- followed by severe abdominal pain and diarrhea

**Treatment:** All three forms of anthrax are treatable with antibiotics. Chances of coming into contact with anthrax are very low, and your body naturally fights off bacteria so you may not even become ill.

**Botulism** - is a muscle-paralyzing disease caused by a toxin made by a bacterium called *Clostridium botulinum*. *C. botulinum* occurs naturally and can be found in soil, water, animals, contaminated foods or crops. According to Health Canada, the toxin produced by *C. botulinum* is the most potent toxin known and can affect humans, animals and even fish. There is only one form of human-made botulism known to date.

**How it spreads:** Botulism <u>cannot</u> spread from person to person. People come into contact with the naturally formed bacteria by eating something (**foodborne** - usually due to improper storage or home canning methods), through a cut in the skin (**wound**), or a small number of infants (typically less than a year old) can eat bacterial spores that get into intestines (**infant botulism**). The only human-made form has been known to be transmitted from monkeys to veterinarians or lab workers (**inhalation**).

**Signs & Symptoms:** Depends on type of botulism you're exposed to and the degree of exposure to the toxin but generally ...

- Foodborne - rare - signs usually appear in 6 to 36 hours
- Wound - first signs usually appear in 4 to 8 days
- Infant botulism - signs usually appear in 6 to 36 hours
- Inhalation - first signs usually appear in 72 hours

Early symptoms for <u>ALL</u> forms of botulism include double vision, blurred vision, drooping eyelids, hard to speak or swallow, dry mouth and fatigue (very tired). Muscle weakness starts at top of body and goes down causing nerve damage that results in paralysis of face, head, throat, chest, arms and legs -- could possibly lead to death since breathing muscles do not work.

**Treatment:** There is an antitoxin for botulism, but it must be treated as quickly as possible since it may or may not reverse the effects of the disease but can stop further paralysis. Antibiotics are not effective against toxins.

**Plague** - is caused by a bacterium called *Yersinia pestis* that affects animals and humans. *Y. pestis* is found in rodents and their fleas in many areas of the world, including the U.S. The bacterium is easily killed by sunlight and drying but could live up to an hour when released into the air depending on weather conditions.

**How it spreads:** There is only one cause of plague but three different types of illness the infection can cause. One type of infection comes from the bite of an infected flea or gets in through a cut in the skin by touching material infected with bacterium (**bubonic**), another can be spread through the air and inhaled (**pneumonic**), and a third type occurs when plague bacteria multiplies in the blood of a person already infected with plague (**septicemic**).

**Signs & Symptoms:** Plague types may occur separately or in combination with each other ... and all start with fever, headache, weakness, chills (possibly puking and diarrhea) usually within 1 to 10 days of being exposed.

- Bubonic - most common - also develop swollen, tender lymph glands (called buboes). Does <u>not</u> spread from person to person.

- Pneumonic - least common but most deadly -- could be used in attack but hard since sunlight kills it. Also get rapidly developing pneumonia with shortness of breath, chest pain, cough, and sometimes bloody or watery spittle. May cause respiratory failure, shock or death. <u>Can</u> be spread person to person through air (inhaling droplets from a cough, sneeze, etc.)

- Septicemic - can occur with either bubonic or pneumonic plague due to bacteria multiplying in blood. Also develop abdominal pain, shock, and bleeding into skin and other organs. Does <u>not</u> spread from person to person.

**Treatment:** There are several antibiotics that can effectively treat plague. (It is very important to get treatment for <u>pneumonic</u> plague within 24 hours of first symptoms to reduce the chance of death).

**Ricin** - is said to be one of the most toxic natural poisons made very easily from the waste left over from processing castor beans. A castor bean plant is a shrub-like herb with clustered seed pods containing bean-like seeds. Accidental poisoning by ricin is unlikely -- it would have to be a planned act to make and use the toxin as a weapon. Ricin can be in many forms and is not weakened much by extreme hot or cold temperatures. (Ricin is also classed as a "biotoxin" under the CDC's chemical agents' list.)

**How it spreads:** Ricin <u>cannot</u> be spread person to person. People come into contact with ricin by breathing in a mist or powder spread into the air (**inhalation**), by eating or drinking something containing toxin (**ingestion**), or by having a ricin solution or pellet stuck into the body (**injection**).

It is hard to say how much ricin could kill a person since it depends on how that person was exposed to the toxin. For example, about 500 micrograms (about the size of the head of a pin) could kill a person if injected into the body, but it would take a lot more if inhaled or swallowed. Ricin prevents cells from making proteins they need when toxin gets inside the body. Cells will die without proteins and eventually the entire body shuts down and dies.

**Signs & Symptoms:** Depends on how much ricin a person is exposed to -- in large amounts death could occur within 36 to 48 hours. If a person lives more than 5 days without problems, there's a good chance they will survive:

- Inhalation - within a few hours of breathing in large amounts of ricin, the first signs are usually coughing, tightness in the chest, hard time breathing, nausea (sick to stomach), and aching muscles. In the next few hours, airways (lungs) would become severely inflamed (swollen and hot), excess fluid would build up in lungs, becomes even harder to breathe, and skin may turn blue

- Ingestion - if a large amount of ricin is swallowed it will cause internal bleeding of the stomach and intestines, leading to puking and bloody diarrhea -- and most likely lead to liver, spleen and kidneys shutting down, and the person could die

- Injection - if enough ricin is injected into a person, it immediately kills muscles and lymph nodes around area where it entered body -- eventually organs would shut down and the person would have massive bleeding from stomach and intestines causing death

**Treatment:** There is no antidote for ricin exposure. Supportive medical care could be given based on how a person was exposed to ricin (like oxygen or medication to reduce swelling [if inhalation] or I.V. fluids [if ingestion]), but care mainly helps symptoms.

Smallpox - is a very serious, highly contagious and sometimes deadly disease caused by the variola virus. The most common form of smallpox causes raised bumps on the face and body of an infected person. There has not been a case of smallpox in the world since 1977, however in the 1980s all countries consolidated their smallpox stocks in two government-controlled laboratories in the U.S. and Russia. These secured laboratories still have the virus in quantities for research purposes, but it is very possible some vials have gotten or could get into the hands of terrorist groups.

Smallpox disease killed over 300 million people in the 20th century and experts say it is the most dangerous infectious disease ever. There is no cure for smallpox and most patients infected with the disease recover, but death may occur in as many as 3 of every 10 persons infected.

**How it spreads:** Smallpox is primarily spread person to person through droplets that are inhaled but usually requires close contact. It can also be spread by infected bodily fluids (especially fluid from bumps) or from bed linens or clothing from an infected person. It is very rare but the virus could carry in the air of an enclosed area like a train or building. Smallpox only infects humans and is not known to be transmitted by insects or animals.

Someone carrying the virus may not even know they have it since it lies dormant (incubation period) for up to 17 days. A person with smallpox is most contagious from the time the rash starts until the last scab falls off (usually about 1 month). Anyone face-to-face with an infected person (within 6 - 7 feet / 2 meters) will most likely get the virus by inhaling droplets or dried fluids or by touching infected materials.

**Signs & Symptoms:** According to the CDC, exposure to the smallpox virus has an incubation period of 7 to 17 days (average is 12 to 14 days) where people feel fine, show no symptoms and are not contagious, then...

- Prodrome phase - first symptoms of smallpox include fever, weird or uneasy feeling (malaise), head and body aches, and sometimes puking. Fever may be high (between 101F-104F or 38C-40C) -- may be contagious. Phase can last 2 - 4 days.

- Spotty mouth - small red spots appear on the tongue and in the mouth (this is start of the "early rash phase")

- Spots become sores - spots turn into sores that break open and spread large amounts of the virus into the mouth and throat -- person is VERY contagious at this point!

- Rash - as sores in mouth break down, a rash starts on the face and spreads to arms and legs, then hands and feet -- usually takes about 24 hours to cover body. As the rash appears, fever drops, and person may feel a little better.

- Raised bumps - by third day, rash turns into raised bumps

- Bumps fill up - by fourth day, bumps fill with a thick, clear fluid -- each bump has a dent in the center (like a bellybutton)

- Bumps become pustules - fever returns and bumps become pustules (which is a raised bump, usually round, firm and feels like there's something hard inside - like a BB pellet) -- lasts about 5 days

- Pustules become scabs - fever still high, next the pustules form a crust turning into scabs - lasts about 5 days -- about 2 weeks after rash first appears most of the sores will be scabbed over

- Scabs fall off - takes about 6 days for all the scabs to fall off leaving a scar or dent in the skin where each scab was (most are gone about 3 weeks after early rash first appears). Person is no longer contagious when all scabs have fallen off.

**Treatment:** There is no cure or treatment for smallpox. A vaccination within 4 days of being exposed could help stop disease but, if vaccinated years ago, it's doubtful you'd be protected now. Many countries are stockpiling vaccine and considering vaccinations for all citizens, but many experts feel that may not be necessary yet. There are certain people who should not get the vaccine. If you do decide to take vaccination, boost your immune system before getting shots since it may help your body fight any adverse reactions.

*If you have concerns or questions about* **smallpox**, *you should visit the CDC's Public Health Emergency Preparedness and Response web site at www.bt.cdc.gov/agent/smallpox or Health Canada's Emergency Preparedness and Response site at www.hc-sc.gc.ca/english/epr/smallpox.html*

<u>Tularemia</u> - (also known as "rabbit fever") is a disease caused by a strong bacterium, *Francisella tularensis*, found in wild animals and some insects (especially rabbits, hares, beavers and other rodents, mosquitoes, deerflys or ticks) and found in soil, water sources and vegetation in those critters' habitats. *F. tularensis* is one of the most infectious bacteria known and it doesn't take much to cause the disease. Tularemia has been considered useful as an airborne weapon worldwide since the 1930s which is why there's valid concern it could be used today in a terrorist attack.

**How it spreads:** Tularemia is not known to spread person to person. Some wild animals carry the disease - usually because they were bitten by an infected bug or drank or ate from contaminated water or soil. Hunters and people who spend a lot of time outdoors can get the disease from critters through a bite or handling a diseased carcass (**skin**), from eating an infected animal not properly cooked or by drinking untreated, contaminated water (**stomach**), or from breathing in dust from contaminated soil (**lungs**).

**Signs & Symptoms:** Depends on how a person is exposed to tularemia and all symptoms may not occur -- all 3 usually appear in 3 to 5 days (but could take up to 14 days) ... may include fever, chills, joint pain, weakness, and ...

- <u>Skin</u> - may also include a bump or ulcers on bite, swollen and painful lymph glands
- <u>Stomach</u> - may also include sore throat, abdominal pain, ulcers on or in mouth, diarrhea or puking
- <u>Lungs</u> - may also include dry cough, chest pain, bloody spittle, trouble breathing or stops breathing

**Treatment:** Tularemia can be treated with antibiotics but people exposed to *F. tularensis* should be treated as soon as possible since it could be deadly.

<u>Viral hemorrhagic</u> [hem-er-á-jik] <u>fevers (VHFs)</u> - are a group of diseases or illnesses caused by several families of viruses. There are many types of VHFs - some the public may recognize are Ebola, Marburg or hantavirus. Some VHFs cause mild reactions or illnesses while others are deadly. Most VHFs are highly contagious and associated with bleeding (hemorrhage), but that's usually not life-threatening. In severe cases, the overall vascular - or blood vessel - system is damaged so the body can't regulate itself thus causing organs to shut down.

Viral hemorrhagic fevers (VHFs) are quite an extensive and complex topic so we are only mentioning it here since it's on the CDC's Category A list. Both the CDC and Health Canada cover VHFs at length on their web sites (listed below) if you would like to learn more. We're just briefly explaining how it can spread and listing some general signs and symptoms in the event you ever hear about "viral hemorrhagic fevers" in the news.

**How it spreads:**  Most viruses associated with VHFs naturally reside in animals (mice, rats or other rodents) or insects (ticks or mosquitoes).  Some VHF viruses could spread to humans by the bite of an infected insect or by breathing in or touching an infected animal's pee, poop, or other body fluids. (For example, a person crawling in a rat-infested area could stir up and breathe in a virus, or someone slaughtering livestock infected by an insect bite could also spread the virus.) Some other VHF viruses spread person to person through direct, close contact with an infected person's body fluids.

**Signs & Symptoms:**  Signs vary by the type of VHF, but the first symptoms often include sudden fever, fatigue (very tired), dizziness, weakness and headache. Person could also have a sore throat, abdominal pain, puking, and diarrhea. Severe cases often show signs of bleeding under the skin, in internal organs, or from the mouth, eyes, or ears.  Blood loss is rarely the cause of death and is usually followed by collapse, shock, coma, seizures and organ failure (the body just shuts down).

**Treatment:**  There is no specific cure or vaccine for most VHFs. Hospitalization and supportive medical care could be given in strict isolation to prevent the virus from spreading to others, but care mainly helps symptoms.  Keeping rodents and mosquitoes out of your home is good prevention.


## BEFORE A BIOLOGICAL ATTACK:

If you skipped the last several pages discussing some BIOLOGICAL agents, you may want to review them along with the following BEFORE, DURING and AFTER tips developed by FEMA and the Department of Homeland Security.

Watch & listen for signs - Many biological agents do not give immediate "warning signs" -- and most symptoms show up hours or days later so it's hard to say what to watch for, but learn about some common agents (see previous pages) and stay current by listening to radio and TV reports to hear what local authorities tell people to do -- and DO it!

Report strange things - Be aware of your surroundings -- watch for strange or suspicious packages ... or spray trucks or crop dusters in weird places at strange times ... and report suspicious activities to local authorities.

<u>Make a plan</u> - Review Section 1 to develop a **Family Emergency Plan** and **Disaster Supplies Kit**.

<u>Get rid of pests</u> - Keep home clean and put food away that might attract rats or mice and get rid of "standing water" sources around yard (like buckets, tires, pots, or kiddie pools) since they are breeding grounds for mosquitoes.

<u>Be ready to evacuate</u> - Listen to local authorities and leave if you are told to evacuate. *(see EVACUATION)*

## DURING A BIOLOGICAL ATTACK:

During any type of biological attack, local authorities will instruct the public about where to go and exactly what to do if exposed to an agent (which may require immediate attention with professional medical staff). It's possible there may be signs (as seen with the anthrax mailings), but more likely it would be discovered <u>after</u> the fact when local health care workers have a wave of sick people seeking emergency medical attention or there are reports of unusual illnesses or symptoms.

<u>Don't panic -- Listen</u> - Stay calm and listen to radio, TV and officials to ...
- Determine if your area is in danger or if you were in the area when it was contaminated.
- Learn signs and symptoms of agent or disease (see previous pages briefly describing **anthrax**, **botulism**, **plague**, **ricin**, **smallpox**, **tulermia**, and **viral hemor-rhagic fevers** [**VHFs** - like Ebola or Marburg] ).
- Find out if medications or vaccines are being distributed by authorities and, if so, where can you get them.

<u>Cover up</u> - Cover your mouth and nose with layers of fabric to filter air but still let you breathe (like 2-3 layers of cotton T-shirt or towel or several layers of paper towel, napkins or tissues).

<u>Clean up</u> - Wash with soap and water to keep from spreading germs.

<u>Stay away</u> - Get away from the attack site to avoid contamination.

<u>Evacuate...?</u> - If you are told to evacuate... DO it!  If officials say you have time, close windows, shut vents and turn off attic fans.  *(see EVACUATION)*

<u>Things to avoid:</u>
- **powder** - strange white powdery substance (anthrax)
- **aerosol mists** - could drift for miles but may be hard to see
- **contaminated food or water** - don't eat or drink any food or water that may have been exposed to agents

<u>Feel sick...?</u> - Many symptoms from biological agents take time to show up so watch family members for signs of illness.

## AFTER A BIOLOGICAL ATTACK:

<u>Don't panic -- Listen</u> - Stay calm and listen to radio, TV and officials to ...
- Determine if your area is or was in danger.
- Learn signs and symptoms of agent or disease.
- Find out if medications or vaccines are being distributed by authorities and, if so, where you can get them.

<u>Feel sick...?</u> - In most cases, people won't be aware they have been exposed to an agent -- some cause immediate symptoms but many take a while to show up so keep watching for signs of illness.

<u>Don't go there</u> - Do not return home until local authorities say it is safe.

<u>Don't spread it</u> - A person, critter, or item that has been exposed to a disease or biological agent may spread it so...
- **clean up** - if your skin or clothing comes in contact with a suspected visible powder or liquid, wash with soap and water to keep from spreading germs
- **store clothes & shoes** - put exposed clothing and shoes in tightly sealed containers without touching other materials and call local authorities to ask how to get rid of them
- **strange symptoms** - if unusual symptoms show up, get to a hospital or medical expert right away!
- **tell people you've been exposed** - tell everyone who comes in contact with you that you may have been exposed to a biological agent
- **land and property** - ask local authorities how to clean up (or ask if it's even necessary)

*For more information about **biological agents**, please visit the Center for Disease Control's Public Health Emergency Preparedness and Response web site at www.bt.cdc.gov or Health Canada's Emergency Preparedness and Response web site at www.hc-sc.gc.ca/english/epr or call the CDC Public Response Hotline at 1-888-246-2675 or 1-888-246-2857 (Español) or 1-866-874-2646 (TTY).*

## CHEMICAL AGENTS

Chemical agents are toxic vapors (gas), sprays (aerosols), liquids or solids that can poison people, animals and the environment. Some compounds or agents do have industrial uses, but many are man-made substances designed, developed and stockpiled as military weapons around the world.

Most chemical agents are difficult to produce and very hard to deliver in large quantities since they scatter so quickly. Most are liquids and some may be odorless and tasteless. They could be inhaled, absorbed into the skin, or swallowed from a contaminated food or water source. Chemical agents can take effect immediately or over several hours or days - and can be deadly if exposed to enough of the agent. If exposed, the best thing you can do is distance yourself from the agent and area and get fresh air.

## What chemical agents could be used in an attack?

According to the CDC, there are several **categories** of chemical agents that could potentially be used in a terrorist attack - some common ones include:

- **Blister Agents / Vesicants** (Sulfur Mustard / Mustard Gas or Lewisite) - primarily cause blisters but can also damage eyes, airways, and digestive system
- **Blood Agents** (Arsine or Cyanide) - gets into the blood stream and prevents cells from absorbing oxygen so cells die
- **Choking / Lung / Pulmonary Agents** (Ammonia or Chlorine) - cause breathing problems and lack of oxygen damages organs
- **Incapacitating Agents** (BZ or LSD) - disrupts central nervous system (knocks you out), causes confusion and slows breathing
- **Nerve Agents** (Sarin, Soman, Tabun or VX) - the most toxic agents -- basically turns "off" the body's ability to stop muscles and glands from twitching (body goes into convulsions). Most agents were originally developed as pesticides / insecticides.

Some other categories include ... **Biotoxins** (like Abrin or Ricin), **Caustics** (Hydrofluoric Acid), **Metals** (Arsenic or Mercury), **Organic Solvents** (Benzene), **Riot Control Agents / Tear Gas** (CS or CN), **Toxic Alcohols** (Ethylene Glycol), and **Vomiting Agents** (Adamsite).

Remember, chemical weapons - or chemical warfare - have been around since World War I ... it's unfortunate we have to even discuss it but try not to let this topic frighten you. As we stated earlier, educate yourselves about the types and where to find more information so you are prepared to react in the event of a chemical threat or attack.

## How could chemical agents be used in an attack?

There are several ways chemical agents could be spread:

- **Vapors / Gas / Aerosols** - spread into the air by a bomb or from aircraft, boats or vehicles -- could spread for miles
- **Liquids** - could be released into the air, water or soil or touched by people or animals
- **Solids** - could be absorbed into water, soil or touched

Some chemical agents can remain in the environment and cause problems long after they are released. Again, keep in mind, both the Center for Disease Control and Environmental Protection Agency are working closely with various Departments of Defense and Energy and many officials around the country to monitor systems and security and develop plans. The same can be said for Health Canada and many other Canadian government agencies.

Again, the CDC asks citizens to not be frightened into thinking they need a gas mask or be concerned about water or food sources. In the event of a public health emergency, officials will tell people what actions need to be taken.

### What are the names of some chemical agents and what can they do?

According to the Center for Disease Control's Public Health Emergency Preparedness and Response web site, there are many types of chemical agents - again too many to list here. We're only mentioning several common agents in alphabetical order, but realize there are many others we are not covering that could potentially be used. Always listen to authorities for instructions in the event of a chemical threat or attack.

**BZ (Incapacitating)** - and other stun agents (LSD, etc.) disrupt the central nervous system causing confusion, short-term memory loss and immobility (means you can't move or are incapacitated).

**How it spreads:** BZ could be released by a bomb or sprayed into the air as an aerosol but has been proven unpredictable if used outdoors.

**Signs & Symptoms:** Depends on how person is exposed to BZ and varies by person -- basically it screws with your nervous system causing confusion, dream-like feelings or strange visions (called hallucinations), dilation of the pupils (means pupils bigger than normal), slurred speech, and loss of motor skills (you can't move). It can also slow down breathing and heart rate.

**Treatment:** BZ is treated with an antidote that reverses symptoms for about an hour. May need repeated doses since the effects can last for hours or days.

**Chlorine (Choking / Lung / Pulmonary)** - is used in industry (to bleach paper or cloth), in water (to kill germs), and in household products. Chlorine can be in the form of a poisonous gas or the gas can be pressurized and cooled into a liquid. When gas comes in contact with moist tissues (eyes, throat, or lungs), an acid is produced that can damage these tissues. Chlorine is not flammable but reacts explosively if mixed with certain liquids.

**How it spreads:** Chlorine could be released into water, food or air. People can be exposed by drinking or eating something contaminated with excess amounts of chlorine or by inhaling the poisonous gas. Chlorine gas is yellow-green, smells like bleach, and stays close to the ground as it spreads.

**Signs & Symptoms:** Depends on how much and how exposed but signs may show up during or right after exposure to dangerous amounts of chlorine:

- Skin - *if gas:* burning pain, redness, and blisters
  *if liquid:* skin white or waxy, numbness, blisters (like frostbite)
- Eyes - burning feeling, blurred vision, watery eyes
- Nose, throat & lungs (respiratory tract) - burning feeling in nose and throat, tightness in chest, coughing, hard time breathing or shortness of breath, fluid builds up in lungs within 2 to 4 hours
- Stomach/gastrointestinal: puking, nausea (sick to stomach)

**Treatment:** - There is no antidote for chlorine exposure -- main things are to remove it from body and seek medical attention as soon as possible.

- First - leave area as quickly as possible
  ... if outdoors - get to high ground (avoid low-lying areas)
  ... if in building - get outside to high ground and upwind
- If inhaled - get fresh air as quickly and calmly as possible
- If on clothing or skin - remove clothes and shoes that are contaminated but don't pull anything over head - cut it off body. If possible, seal clothing in plastic bag, then seal that bag in a bag. Immediately wash body with clean water and soap.
- If in eyes - remove contacts and put in bags with clothing - do not put back in eyes! If eyes burning or vision blurred, rinse eyes with plain water for 10 to 15 minutes. If wearing glasses, wash them with soap and water before putting back on.
- If swallowed - if someone drinks or eats something exposed to chlorine, do NOT make them puke or drink fluids - call 9-1-1

Ask officials how to dispose of bags with contaminated clothing, shoes, etc.

**Cyanide (Blood)** - is a very fast acting and potentially deadly chemical that exists in several forms. The CDC categorizes cyanide as "blood" agents but sometimes called "cyanide or cyanogen" agents. Cyanide can be a colorless gas (cyanogen chloride or hydrogen cyanide) or a crystal solid form (like potassium or sodium cyanide). It may smell like "bitter almonds" but most often is odorless. Cyanide is naturally present in some foods or plants - it's also in cigarette smoke or given off when some plastics burn. It is also used to make paper or textiles and in chemicals used to develop photos.

**How it spreads:** Cyanide could enter water, soil or air as a result of natural or industrial processes or be spread indoors or outdoors as a weapon. People can be exposed by breathing gas or vapors or cigarette smoke, by drinking or eating something contaminated (either accidentally or on purpose) or by touching soil or clothing that was exposed to cyanide. Cyanide gas disappears quickly and rises since less dense than air so pretty useless outdoors.

**Signs & Symptoms:** Basically cyanide prevents the cells from absorbing oxygen so cells die. No matter how exposed (breathing, absorbed through skin, or eating / drinking) some or all signs show up within minutes:

- Exposed to small amount - rapid breathing, gasping for air, dizziness, weakness, headache, nausea (sick to stomach), puking, restlessness, rapid heart rate, bluish skin or lips (due to lack of oxygen in blood)

- Large amount - above signs plus convulsions, low blood pressure, slow heart rate, pass out, lung injury, stops breathing leading to death. Survivors of serious poisoning may develop heart and brain damage due to lack of oxygen.

**Treatment:** Cyanide poisoning is treated with antidotes and supportive medical care (mainly to help symptoms). The main things are to avoid area where it was released and seek medical attention as soon as possible.

- First - leave area as quickly as possible
  ... if outdoors - move upwind or stay low to ground (gas rises)
  ... if in building - get outside and get upwind
- If inhaled - get fresh air as quickly and calmly as possible
- If on clothing or skin - remove clothes and shoes that are contaminated. Seal clothing in plastic bag then seal that bag in a bag - ask how to dispose of bags. Immediately wash any exposed body parts (skin & hair) with clean water and soap.
- If in eyes - remove contacts if necessary. If eyes are burning or vision blurred, rinse eyes with plain water for 10 to 15 minutes.
- If swallowed - if someone drinks or eats something exposed to cyanide, do NOT make them puke or drink fluids - call 9-1-1

Sarin (**Nerve**) - is a clear, colorless, odorless and tasteless liquid that could evaporate into a vapor (gas) and contaminate the environment. It is man-made and originally developed to kill insects. Nerve agents basically turn "off" the body's ability to stop muscles and glands from twitching.

**How it spreads:** Sarin could be released into the air, water, or soil as a weapon. People can be exposed by breathing vapors, by drinking or eating something contaminated, or by touching water, soil or clothing exposed to sarin. A person's clothing can release sarin for about 30 minutes after being exposed to vapor. Because sarin vapor is heavier than air, it settles in low-lying areas creating a greater exposure hazard.

**Signs & Symptoms:** Depends on how much, what form, and how people are exposed to sarin. No matter how exposed (breathing, absorbed through skin, or eating / drinking it), the following may show up within seconds (vapor or gas) or within minutes to 18 hours (liquid)...

- Head - runny nose, drooling or excess spittle, headache
- Eyes - watery, small pupils, blurred vision, eye pain

- **Lungs** - cough, tight feeling in chest, fast / rapid breathing
- **Nervous system** - confusion, drowsiness, weakness
- **Heart/blood** - slow/fast heart rate, rise/drop in blood pressure
- **Stomach/gastrointestinal** - abdominal pain, puking, nausea (sick to stomach), diarrhea, pee lot more than normal

... plus ...

- **If exposed to small amount** - just a drop of sarin on skin can cause sweating and muscle twitching
- **If large amount** - can cause convulsions (body can't stop the muscles and glands from twitching), paralysis (can't move), pass out, stops breathing leading to death

**Treatment:** Sarin poisoning is treated with antidotes and supportive medical care. Main things are to avoid area where released, get decontaminated (strip & wash), and seek medical attention as soon as possible.

- **First - leave area as quickly as possible**
  ... if outdoors - move to higher ground and stay upwind
  ... if in building - get outside to highest ground possible
- **If inhaled** - get fresh air as quickly and calmly as possible
- **If on clothing or skin** - remove clothes and shoes that are contaminated but don't pull anything over head - cut it off body. Seal all in plastic bag, then seal that bag in a bag and ask how to dispose of. Immediately wash body with clean water and soap.
- **If in eyes** - remove contacts if necessary. If eyes burning or vision blurred, rinse eyes with plain water for 10 to 15 minutes.
- **If swallowed** - if someone drinks or eats something exposed to sarin, do NOT make them puke or drink fluids - call 9-1-1

**Sulfur Mustard / Mustard gas (Blister/Vesicant)** - (also known as mustard agent) can be in the form of a vapor, an oily-textured liquid or a solid and be clear to yellow or brown when in liquid or solid form. It is not normally found in the environment, however, if released, can last for weeks or months under very cold conditions. Under normal weather conditions, it usually only lasts a day or two.

Mustard gas is fairly easy to develop so many countries that decide to have chemical warfare agents usually stock up on this one. Sulfur mustard was originally produced in the 1800's but first used as chemical warfare in World War I and in many wars since. Exposure to sulfur mustard is usually not fatal but could have long-term health effects.

**How it spreads:** Sulfur mustard / mustard gas can be released into the air as a vapor or gas and enter a person's body by breathing or get on skin or in eyes. The vapor would be carried for long distances by wind so the agent could affect a wide area. Sulfur mustard is heavier than air so vapors will settle in low-lying areas. A liquid or solid form could be released into water and a person could be exposed by drinking it or absorbing it through the skin. Since it often has no smell or the smell doesn't raise a red flag (can smell like garlic, onions or mustard), people may not realize they have been exposed.

**Signs & Symptoms:** Depends on how much, what form, and how a person is exposed to sulfur mustard / mustard gas and may not occur for 2 to 24 hours ... some immediate signs include...

- Skin - redness and itching of skin may occur 2 to 48 hours after exposure -- changes to yellow blistering of skin
- Eyes - a mild case causes irritation, pain, swelling and watery eyes within 3 to 12 hours -- a more severe case causes same within 1 to 2 hours - may also include light sensitivity, severe pain or temporary blindness (lasting up to 10 days)
- Nose & lungs (respiratory tract) - runny nose, sneezing, sinus pain, bloody nose, short of breath, may get hoarse, and cough (mild exposure shows within 12 to 24 hours -- severe shows within 2 to 4 hours)
- Digestive tract - abdominal pain, nausea (sick to stomach), diarrhea, puking, and fever

Some long-term health effects may include ...

- Burns or scarring - exposure to sulfur mustard liquid (not gas) may produce second- and third-degree burns and later scarring
- Breathing problems or disease - severe exposure could cause chronic respiratory disease, repeated infections, or death
- Blindness - severe exposure can cause permanent blindness
- Cancer - may increase chance of lung or respiratory cancer

**Treatment:** There is no antidote for sulfur mustard / mustard gas exposure - the best thing to do is avoid it by leaving the area where it was released.

- First - leave area as quickly as possible
  ... if outdoors - move upwind and get to higher ground
  ... if in building - get outside, upwind and to higher ground
- If inhaled - get fresh air as quickly and calmly as possible
- If on clothing or skin - remove everything that got contaminated. Seal clothing and shoes in plastic bag, then seal that bag in a bag - ask how to dispose of later. Immediately wash exposed body parts (eyes, skin, hair, etc.) with plain, clean water.
- If in eyes - remove contacts if necessary. Flush eyes with water for 5 to 10 minutes but do NOT cover eyes with bandages - put on shades or goggles to protect them.
- If swallowed - if someone drinks or eats something contamin-ated with sulfur mustard, do NOT make them puke it up -- give the person some milk to drink and call 9-1-1

<u>VX</u> (**Nerve**) - is an oily liquid that is odorless, tasteless, amber or honey-yellow in color, and evaporates about as slowly as motor oil. VX is the most potent of all nerve agents, which basically turn "off" the body's ability to stop muscles and glands from twitching. Like other nerve agents, VX is a man-made chemical originally developed to kill insects and pests.

**How it spreads:** VX could be released into the air or water as a weapon, however it does not mix with water as well as other nerve agents. If VX gas or vapors are released into the air, people can be exposed by breathing or eye or skin contact and a person's clothing can release VX for about 30 minutes after being exposed. If VX liquid is put in food or water source, people could get it from eating, drinking or touching something exposed to the liquid.

VX vapor is heavier than air so settles in low-lying areas. Under average weather conditions, VX can last for days on objects that come in contact with the agent, but in cold weather it could last for months. The liquid takes time to evaporate into a vapor so could be a long-term threat to the environment.

**Signs & Symptoms:** VX is similar to sarin - depends on how much, what form, and how people are exposed. No matter how exposed (breathing, absorbed through skin, or eating / drinking it) the following may show up within seconds to hours ...

- <u>Head</u> - runny nose, drooling or excess spittle, headache
- <u>Eyes</u> - watery, small pupils, blurred vision, eye pain
- <u>Lungs</u> - cough, tight feeling in chest, fast / rapid breathing
- <u>Nervous system</u> - confusion, drowsiness, weakness
- <u>Heart/blood</u> - slow or fast heart rate, rise or drop in blood pressure
- <u>Stomach/gastrointestinal</u> - abdominal pain, puking, nausea (sick to stomach), diarrhea, pee more than normal

... plus ...

- <u>If exposed to small amount</u> - a tiny drop of VX on skin can cause sweating and muscle twitching
- <u>If large amount</u> - can cause convulsions (body can't stop the muscles and glands from twitching), paralysis (can't move), may pass out, stops breathing leading to death

**Treatment:** VX poisoning can be treated with antidotes but must be given shortly after exposure to be effective. The main things are avoid area where agent was released, get decontaminated (strip & wash), and seek medical care as soon as possible.

- <u>First - leave area as quickly as possible</u>
  ... if outdoors - move to higher ground and stay upwind
  ... if in building - get outside to highest ground possible
- <u>If inhaled</u> - get fresh air as quickly and calmly as possible

- If on clothing or skin - remove clothes and shoes contaminated with VX but don't pull anything over head - cut it off body. Seal all in plastic bag, then seal that bag in a bag and ask how to dispose of. Immediately wash body with clean water and soap.
- If in eyes - remove contacts if necessary. If eyes burning or vision blurred, rinse eyes with plain water for 10 to 15 minutes.
- If swallowed - if someone drinks or eats something exposed to VX, do NOT make them puke or drink fluids - call 9-1-1

## BEFORE A CHEMICAL ATTACK:

If you skipped the last several pages discussing some CHEMICAL agents, you may want to review them, along with the below BEFORE, DURING and AFTER tips developed by FEMA and the Department of Homeland Security. You may also want to review the HAZARDOUS MATERIALS topic.

Watch & listen for signs - Many chemical agents can cause watery eyes, choking, trouble breathing, coughing or twitching. If you see or hear a lot of people doing this or see a bunch of birds, fish or small animals sick or dead, it should raise a red flag. Learn about some common potentially hazardous chemical agents (see previous pages) and stay current by listening to radio and TV to hear what local authorities tell people to do -- and DO it!

Report strange things - Be aware of your surroundings -- watch for strange or suspicious packages ... or spray trucks or crop dusters in weird places at strange times ... and report suspicious activities to local authorities.

Make a plan - Review Section 1 to develop a **Family Emergency Plan** and **Disaster Supplies Kit**. Some key items include a battery-powered radio (with extra batteries), food and drinking water, duct tape, plastic and scissors, first aid kit, and sanitation items (soap, extra water and bleach).

Pick a room - It could take authorities time to determine what (if any) agent was used so pick a room in advance your family could use if told to stay indoors for several hours. It's best to pick an internal room where you could block out air IF told to do so. To save time consider measuring and cutting plastic sheets in advance for openings (vents, windows, and doors). Remember, a toilet may be vented meaning outside air comes in constantly or when flushed (depends on design) - in case using bathroom as a safe room.

Calculate air for room - Keep in mind people can stay in a sealed off room for only so long (or you'll run out of air!) FEMA suggests 10 square feet of floor space per person (like 5ft x 2ft / 1.5m x 0.6m ) will provide enough air to prevent carbon dioxide buildup for up to 5 hours.

Be ready to evacuate - Listen to local authorities and leave if you are told to evacuate. *(see EVACUATION)*

# DURING A CHEMICAL ATTACK:

During any type of chemical attack, local authorities will instruct the public on where to go and exactly what to do if exposed to an agent (which may require immediate attention with professional medical staff).

Watch for signs - If you see or hear a lot of people choking, coughing or twitching or see a bunch of sick or dead critters - leave area quickly!

Don't panic -- Listen - Stay calm and listen to radio, TV and officials to ...
- Determine if your area is or was in danger.
- Learn signs and symptoms of some agents (see previous pages briefly describing **BZ**, **chlorine, cyanide, sarin, sulfur mustard / mustard gas**, and **VX**).
- Find out if and where antidotes are being distributed.

IF INDOORS – Stay inside and ...
- Close your windows, vents and fireplace damper and turn off A/C and fans to reduce air drawn in from outside.
- Seal gaps under doorways and windows with wet towels, plastic (if available) and duct tape.
- If you picked a safe room in advance, grab your **Disaster Supplies Kit** and seal off that room - remember, you can only stay there for so many hours or you'll run out of air.

IF OUTDOORS - Stay upwind from the disaster area since many agents can be carried by wind. Try to find a shelter as quickly as possible!

IF IN A VEHICLE - Close your windows and shut off vents to reduce risk and drive away and upwind from the attack site, if possible.

Cover up - Cover your mouth and nose with layers of fabric to filter air but still let you breathe (like 2-3 layers of cotton T-shirt or towel or several    layers of paper towel, napkins or tissues).

Feel sick...? - Some agents can cause immediate symptoms and some take a while to show up so watch family members for signs of illness.

Evacuate...? - If you are told to evacuate… DO it!  If officials say you have time, close windows, shut vents and turn off attic fans. *(see EVACUATION)*

Things to avoid:
- **chemicals** - spilled liquid materials or vapors or gas
- **contaminated food or water** - don't eat or drink any food or water that may have been exposed to materials

<u>Stay away</u> - Get away from the attack site to avoid contamination.

## AFTER A CHEMICAL ATTACK:

<u>Feel sick...?</u> - In some cases, people won't be aware they have been exposed to an agent -- most cause immediate symptoms and some take a while to show up so continue watching for signs of illness.

<u>Don't panic -- Listen</u> - Stay calm and listen to radio, TV and officials to ...
- Determine if your area is or was in danger.
- Learn signs and symptoms of chemical agent.
- Find out if antidotes are being distributed by authorities and, if so, where can you get them.

<u>Don't go there</u> - Don't return home until local authorities say it is safe.

<u>Air out</u> - Open windows, vents and turn on fans in your home.

<u>Clean up</u> - A person, critter or item that has been exposed could spread it...
- **decontamination** - follow instructions from local authorities since it depends on the chemical. May need to shower with or without soap or may be told to avoid water - check first!
- **strange symptoms** - if unusual symptoms show up, get to a hospital or medical expert right away!
- **store clothes & shoes** - put exposed clothing and shoes in tightly sealed containers without touching other materials and call local authorities to ask how to get rid of them
- **tell people you've been exposed** - tell everyone who comes in contact with you that you may have been exposed to a chemical agent
- **land and property** - ask local authorities how to clean it up

<u>Strange vapors or danger</u> - Report any strange vapors or other dangers to the local authorities immediately.

*For more information about **chemical agents**, please visit the Center for Disease Control's Public Health Emergency Preparedness and Response web site at www.bt.cdc.gov or call the CDC's Public Response Hotline at 1-888-246-2675 or 1-888-246-2857 (Español) or 1-866-874-2646 (TTY).*

## RADIOLOGICAL THREAT OR DEVICE
Due to the heightened threat of terrorist attacks using a "dirty bomb", we are briefly covering it here. In most cases, terrorist attacks would <u>not</u> involve a

nuclear device (like a "dirty nuke" or missile) since they require weapons grade uranium or plutonium which are very difficult to obtain and develop, but listen for alerts and advisories from officials.

*Please note, we already covered nuclear accidents and what to do BEFORE, DURING or AFTER a nuclear-related incident (see pages 74-78).*

## What is a "dirty bomb"?

A radiological dispersion device (**RDD**) - also known as a "dirty bomb" - uses conventional explosives (like dynamite) to spread radioactive materials in the form of powder or pellets over a targeted area.

This type of attack appeals to terrorists since it doesn't require a lot of technical know-how to build and use ... plus low-level radioactive materials are pretty easy to obtain since they are used in many fields like agriculture, research and medicine.

The most harmful, high-level radioactive materials would be found in nuclear power plants and at nuclear weapons sites, but with the heightened state of alert at many of these locations, it'd be very dangerous and difficult for terrorist organizations to get them.

## What are the dangers of an RDD ("dirty bomb")?

A terrorist's main reasons for using a "dirty bomb" is to cause damage to buildings, contaminate an area, and spread fear or panic.

According to the Center for Disease Control, the primary danger from a dirty bomb would be the blast itself - not necessarily the radiation. Knowing how much (if any) radiation might be present at the attack site is difficult when the source of the radiation is unknown until site is tested. However, since many RDDs could be using low-level radioactive materials, there probably would not be enough radiation to cause severe illness.

## Has anyone used a "dirty bomb" before?

According to a United Nations report, Iraq tested a dirty bomb device in 1987 but found the radiation levels were too low to cause significant damage. Thus, Iraq abandoned any further use of the device.[5]

## What if you or your office receives a "bomb threat"?

Bomb threats are usually received by a telephone call or in the mail. It is highly unlikely a terrorist organization using a "dirty bomb" would give anyone advance warning or call with a bomb threat, however, in the event you or someone in your office receives a bomb threat, do the following...

- If you ever receive a bomb threat, get as much information from the caller as possible.

- Try to keep caller on the phone as long as you can and write down <u>everything</u> that is said! (Since you'll be nervous or scared, good notes will be very helpful to officials later!)
- Notify the police and building management.
- Calmly evacuate the building, keep the sidewalks clear and stay away from windows.

### <u>What if you or someone in your office receives a "suspicious package"?</u>

According to the United States Postal Service, the likelihood of you ever receiving a bomb in the mail is remote. Unfortunately, there have been a small number of explosive devices and biological agents that have surfaced in the mail over the years. Some possible motives for an individual or group sending a "suspicious package" include revenge, extortion, love triangles, terrorism, and business disputes.

The following are some unique signs or characteristics from the U.S. Postal Inspection Service that may help identify a "suspect" piece of mail ...

- Package may have restricted markings like "Personal" or "Private" to one who doesn't receive personal mail at office.
- Package is sealed with excessive amounts of tape or has way too much postage on it.
- Postmark city different than Return Address city.
- Misspelled words, written badly or done with letters cut from newspaper or magazine and pasted on.
- Package has wires or aluminum foil sticking out, oil stains, smells weird or sounds funny (sloshing noise).
- Package may feel strange or look uneven or lopsided.

If you are unsure about a letter or package and are not able to verify the Sender or contents with the person it is addressed to then...

- DO NOT open it, shake it, bump it or sniff it!
- Cover the letter or package with a shirt, trash can or whatever is handy.
- Evacuate the area quickly and calmly.
- Wash your hands with lots of soap and water.
- Call building security, police and your postal inspector.
- List all the people who were near the package or letter in case they are needed for further questioning.

## BEFORE A RADIOLOGICAL THREAT OR EVENT:

<u>Make a plan</u> - Review Section 1 and check on emergency plans for schools, day cares, nursing homes, etc. and where everyone goes when / if evacuated.

Be ready to evacuate - Listen to authorities -- if told to leave - DO it!

Learn about radiation - Please review pages 74-78 since a nuclear-related incident is a similar situation.

## DURING A RADIOLOGICAL EVENT OR EXPLOSION:

Don't panic... - Get out of the area as quickly and calmly as possible!

Don't look... - Do NOT look directly at explosion, flash, blast or fireball!

Things to watch out for:
- **falling objects** - if things are falling off bookshelves or from the ceiling, get under a sturdy table or desk
- **fire** - stay low to the floor (crawl or walk like a duck)
  - only use the stairs (don't use elevators)
  - Check doors before opening with back of hand (If HOT, do NOT open - find another exit!)
- **weak floors or stairs** - be careful since floors and stairs could have been weakened by the blast

Cover up - Cover your mouth and nose with layers of fabric to filter air but still let you breathe (like shirt, paper towel, napkins or tissues).

EMP...? - Blast could create an electromagnetic pulse (zap) that can fry electronics connected to wires or antennas like cell phones, computers, cars, etc.

## AFTER A RADIOLOGICAL EVENT OR EXPLOSION:

If you are trapped in an area:
- **light** - use flashlight – never use matches or lighters in case there are gas leaks
- **be still** - try to stay still so you won't kick up dust
- **breathing** - cover your mouth with a piece of clothing
- **make noise** - tap on a pipe or wall so rescuers can hear you (shout only as a last resort -- you could inhale too much dust)

Get distance & shielding - Get out of area quickly and into nearest building to reduce chances of being exposed to radioactive materials (if any).

Rescuing others - Untrained persons should not try to rescue people who are inside a collapsed building… wait for emergency personnel to arrive – then, IF they need you, they will ask.

**If radioactive materials were possibly present:**

Don't panic -- Listen - Stay calm and listen to radio, TV and officials to ...
- Determine if your area is in danger.
- Find out where to go for radiation monitoring and blood tests to determine if exposed and what to do to protect health.
- Learn if **KI** (potassium iodide) is being passed out by authorities and, if so, find out where to get tablets... or ask if you should take them (if in your **Disaster Supplies Kit**).

Will I get radiation sickness or cancer...? - In most cases, people won't be aware they have been exposed since radiation cannot be seen, smelled, felt, or tasted. Just because you were at the site of a dirty bomb does not mean you were exposed to radioactive material. Until doctors are able to check skin with sensitive radiation detection devices or run blood tests to determine there was any radiation - no one really knows if they were exposed. And even if you were exposed to small amounts of radioactive material, it does not mean you will be sick or get cancer. Listen to and work with medical health professionals since it depends on each specific situation or incident.

*For the CDC's information on "acute radiation syndrome" (radiation sickness) visit  www.bt.cdc.gov/radiation/ars.asp*

*For more information about **radiological emergencies** or **radiation emergencies**, please visit the Center for Disease Control's Public Health Emergency Preparedness and Response web site at  www.bt.cdc.gov  or visit Health Canada's Healthy Environments and Consumer Safety Radiation Protection Bureau at  www.hc-sc.gc.ca/hecs-sesc/rpb*


## WEAPONS OF MASS DESTRUCTION (WMD)

The EPA's Counter Terrorism site describes WMD as "weapons or devices that are intended, or have the capability, to cause death or serious bodily injury to a significant number of people, through the release, dissemination, or impact of toxic poisonous chemicals; disease organisms; or radiation or radioactivity." Recent events indicate several nations with WMDs have ties or suspected ties to terrorist groups so the threat of an attack is very real.

In the event of a threat against North America with WMD, there are officials from all levels of government responsible for employing and equipping WMD terrorism response units to manage the situation. The information in this book may be helpful, but it is critical the general public stays calm and listens to local and federal officials for specific instructions.

*You may want to review APPENDIX C "County Emergency Preparedness Terrorism Emergency Operations Outline" contributed by South Carolina's Charleston County Emergency Preparedness Department.*

# What are <u>YOU</u> gonna do about...
## A Thunderstorm?

Thunderstorms are very common... in fact, at any given moment, nearly 1,800 thunderstorms can be in progress over the face of the earth!  The U.S. usually averages about 100,000 thunderstorms each year.

Lightning always comes with a thunderstorm since that is what causes the thunder.  If you have ever heard someone say lightning never strikes the same place twice... WRONG... it can!  In fact, lightning OFTEN strikes the same place several times during one storm.  Lightning actually comes from the ground up into the air and back down - we just see it as it comes down so it looks like it's coming from the clouds.

Severe thunderstorms can also bring heavy rains, flooding, hail, strong winds, tornadoes and microbursts (a sudden vertical drop of air).

## Before a Thunderstorm:

<u>Prepare</u> - Review WIND, FLOOD, and LIGHTNING MITIGATION at beginning of this Section.

<u>Learn the buzzwords</u> - **Learn the terms / words used with thunderstorms...**
- **Severe Thunderstorm Watch** - tells you when and where severe thunderstorms are possible
- **Severe Thunderstorm Warning** - severe thunderstorms are have been spotted or are occurring

<u>Watch for lightning</u> - If you hear thunder, you're close enough to be struck by lightning - take cover as quickly as possible.

<u>Unplug it</u> - Unplug appliances if possible - even ones on a surge protector and it's best to move plugs away from outlets.

## During a Thunderstorm:

<u>Listen</u> - Keep a battery-operated radio near by for local reports on storm (especially severe storms which could cause tornadoes!)

**IF INDOORS** - Stay inside until the storm passes.
- <u>Don't shower</u> - sounds weird, but it's best to avoid taking a bath or shower since water can carry an electrical charge if lightning strikes near your home

- Telephone - best not to use since phone lines can conduct electricity (could shock you).

**IF OUTDOORS** - Try to get to safe shelter quickly.
- Move away from tall things (trees, towers, fences, telephone or power lines) and metal things (umbrellas, motorcycles or bicycles, wire fences, etc) since they all attract lightning.
- If surrounded by trees, take shelter under the shorter trees.
- Get to a low lying area (like a ditch or a valley) but watch out for flash floods.
- **Be small** - make yourself a small target by crouching down and put your hands on your knees (and don't lie flat on the ground since that makes you a bigger target!)

**IF IN A BOAT** - Get to land and to shelter quickly! Water is extremely dangerous when there's lightning!

**IF IN A VEHICLE** - Keep windows closed and stay out of a convertible, if possible (mainly because the top is usually fabric and that could make YOU the highest target if lightning strikes).

Hairy sign - If you feel your hair stand on end and feel tingly (which means lightning is about to strike)… crouch down and bend forward putting hands on your knees (be small)! Don't lie flat on ground… makes a bigger target!

If someone is struck by lightning:
- Victim does not carry electrical charge and CAN be touched.
- Call 9-1-1 or your local EMS (emergency) telephone number.
- Victim will have 2 wounds - an entrance and an exit burn and give first aid, if needed. *(see BURNS [Electrical] in Section 3)*

## AFTER A THUNDERSTORM:

Things to avoid:
- **flooded areas** – stay away from flood waters since it may be contaminated by oil, gasoline or raw sewage or electrically charged from underground or downed power lines or lightning – wait for authorities to approve returning to flooded areas
- **moving water** – 6 inches (15 cm) of moving water can knock you off your feet and 2 ft (.6 m) of moving water can float a car
- **storm-damaged areas**
- **downed power lines**

Recovery tips - Review TIPS ON RECOVERING FROM A DISASTER at end of this Section.

# What are <u>YOU</u> gonna do about...
## A TORNADO?

The U.S. has more tornadoes than anywhere else in the world (averaging about 1,000 per year), with sightings in all 50 states. Canada is # 2 in volume of tornadoes (averaging about 80 per year) with several high risk areas mostly in central provinces.

Most injuries or deaths caused by tornadoes are from collapsing buildings, flying objects, or trying to outrun a twister in a vehicle. Tornadoes can also produce violent winds, hail, lightning, rain and flooding.

Dr. T. Fujita developed a damage scale **(Fujita Scale or F-scale)** based on wind speeds and damage potential. *(Note, wind speeds are estimates and not scientifically verified, but this is the most common rating method per NOAA.)*

| Scale | Wind Estimate | Typical Damage *(per NOAA Storm Prediction Ctr)* |
|-------|---------------|--------------------------------------------------|
| F0 | < 73 mph<br>< 117 km/h | **Light**: Some damage to chimneys, signs and vegetation. |
| F1 | 73-112 mph<br>117-180 km/h | **Moderate**: Peels surface off roofs; blows most mobile homes and moving autos around, etc. |
| F2 | 113-157 mph<br>181-253 km/h | **Considerable**: Roofs & mobile homes destroyed; trees snap; light-object missiles generated, etc. |
| F3 | 158-206 mph<br>254-331 km/h | **Severe**: Roofs & walls ripped off sturdy homes; trees uprooted; heavy cars lifted and thrown, etc. |
| F4 | 207-260 mph<br>333-418 km/h | **Devastating**: Most homes leveled; some pieces blown; cars thrown and large missiles generated. |
| F5 | 261-318 mph<br>420-511 km/h | **Incredible**: All homes leveled and swept away; car-sized missiles fly thru air over 100 metres (109 yards); trees debarked & other weird stuff. |

**Did you know...**
> ... the force of a tornado can strip asphalt chunks off roads, rip clothes off people, and pluck feathers off chickens?!

## BEFORE A TORNADO:

<u>Prepare</u> - Review WIND, FLOOD, and LIGHTNING MITIGATION at beginning of this Section.

<u>Learn the buzzwords</u> - Learn the terms / words used with tornado threats...
- **Tornado watch** - a tornado is possible so listen to TV or radio for updates

- **Tornado warning** - a tornado has been sighted so take shelter quickly and keep a battery-operated radio with you for updates

Learn risks - Ask the local emergency management office about threats in your area, what the warning signals are, and what to do when you hear them.

Where am I? - Make sure everyone knows what county or area you live in and listen for that name on radio updates.

Get tuned in - Make sure you have a battery-operated radio (with spare batteries) for weather forecasts and updates. (Some radios like Environment Canada's Weatheradio and NOAA's Weather Radio have a tone-alert feature that automatically alerts you when a Watch or Warning has been issued.)

Be ready to evacuate - Listen to local authorities and if told to leave - do it! *(see EVACUATION)*

Make a plan - Review Section 1 to develop a **Family Emergency Plan** and **Disaster Supplies Kit**.

Learn to shut off - Know where and how to shut off electricity, gas and water at main switches and valves -- ask local utilities for instructions.

Where do I go? - Know locations of shelters where you spend time (schools, nursing homes, office, etc.) The best place is underground (like a basement, a safe room, or storm cellar) or find a hallway, bathroom, or closet in middle of building on the lowest floor.

Do drills - Practice going to shelter with your family and "duck and cover" (use your hands and arms to protect head and stay down low).

Put it on film - Either videotape or take pictures of your home and personal belongings and store them in a safe place (like a fireproof box or a safety deposit box) along with important papers.

## DURING A TORNADO WATCH OR WARNING:

Review above tips and...

Listen - Keep up with local news reports tracking the twister or conditions using a battery-operated radio.

Watch & listen - Some danger signs of a tornado include dark green-ish sky, clouds moving to form a funnel, large hail, or loud roar (like a freight train).

Be ready to evacuate - Keep listening to authorities and leave if told to do so.

# During a Tornado:

<u>Listen</u> - Use a battery-operated radio to hear reports tracking the twister.

<u>Take cover</u> - If you hear or see a tornado coming take cover immediately!

**IF IN A TRAILER OR MOBILE HOME – GET OUT!!!**
- Get to a stronger shelter… or …
- Stay low to ground in a dry ditch and cover head with hands.
- If you hear or see water in the ditch, move quickly to a drier spot (in case lightning strikes nearby).

**IF INDOORS** - Get to a safe place right away - and avoid windows!!
- <u>In house or small building</u> - go to basement, storm cellar or middle of building on lowest floor (a bathroom, closet or hallway). Get under something sturdy or put mattress or covers over you for protection and stay put until danger has passed!
- <u>In a school, nursing home, hospital, factory or shopping center</u> - go to designated shelter areas (or interior hallways on lowest floor) -- stay away from open areas.
- <u>In a high-rise building</u> - go to a small, interior room or hallway on lowest floor possible and avoid windows.

**IF OUTDOORS** - Try to take shelter in a basement or sturdy building!  Or lie in a dry ditch with hands covering your head, but watch and listen for flooding and be aware you're a bigger target for lightning.  (And if you hear or see water, move since it can carry lightning's electrical charge!)

**IF IN A VEHICLE** - GET OUT and take shelter in a building or lie flat in a ditch with hands covering head (but be aware you're a bigger target for lightning when lying flat & listen for flooding!)  DO NOT try to out-drive a tornado! You never know which direction one will go and it moves too fast.


# After a Tornado:

<u>Watch out</u> - Look for broken glass and downed power lines.

<u>Injured people</u> - Do not try to move injured people unless they are in danger and call for help immediately.  *(see TIPS ON BASIC FIRST AID)*

<u>Don't go there</u> - Stay out of damaged buildings or homes until OK'd to enter.

<u>What to wear</u> - Use sturdy work boots and gloves.

<u>Recovery tips</u> - See TIPS ON RECOVERING FROM A DISASTER at end of this Section.

# What are <u>YOU</u> gonna do about...
## A Tsunami?

A tsunami [soo-nah´-mee] is a series of huge, destructive waves caused by an undersea disturbance from an earthquake, volcano, landslide, or even a meteorite. As the waves approach the shallow coastal waters, they appear normal and the speed decreases. Then, as the tsunami nears the coastline, it turns into a gigantic, forceful wall of water that smashes into the shore with speeds exceeding 600 miles per hour (965 km/h)! Usually tsunamis are about 20 feet (6 m) high but extreme ones can get as high as 100 feet (30 m)!

A tsunami is a series of waves and the first wave may <u>not</u> be the largest one, plus the danger can last for many hours after the first wave hits. During the past 100 years, more than 200 tsunamis have been recorded in the Pacific Ocean due to earthquakes and Japan has suffered a majority of them.

**Did you know...**

 ... a tsunami is <u>not</u> a tidal wave - it has nothing to do with the tide?!

 ... another name used to describe a tsunami is "harbor wave" - "tsu" means harbor and "nami" means wave in Japanese?!

 ... sometimes the ocean floor is exposed near the shore since a tsunami can cause the water to recede or move back before slamming in to shore?!

 ... boats, rocks and other debris can be moved inland hundreds of feet with power that can destroy everything in its' path?!

 ... tsunamis can travel up streams and rivers that lead to ocean?!

## Before a Tsunami:

<u>Prepare</u> - Review WIND and FLOOD MITIGATION at beginning of this Section.

<u>Learn the buzzwords</u> - Learn the words used by both the West Coast / Alaska Tsunami Warning Center (WC/ATWC - for AK, BC, CA, OR, and WA) and the Pacific Tsunami Warning Center (PTWC - for international authorities, HI and all U.S. territories within Pacific basin) for tsunami threats...

- **Advisory** - an earthquake has occurred in the Pacific basin which might generate a tsunami
- **Watch** - a tsunami was or may have been generated, but is at least 2 hours travel time from Watch area
- **Warning** - a tsunami was or may have been generated and could cause damage to Warning area - should evacuate

Learn risks - If new to area, call local emergency management office about tsunami threat and learn what warning signals are and what to do when you hear them.

Make a plan - Review Section 1 to develop a **Family Emergency Plan** and **Disaster Supplies Kit**.

Listen - Make sure you have a battery-operated radio (with spare batteries) for weather forecasts and updates. (Radios like Environment Canada's Weatheradio and NOAA's Weather Radio have a tone-alert feature that automatically alerts you when a Watch or Warning has been issued.)

Water signs - If near water or shore, watch for a noticeable rise or fall in the normal depth of coastal water - that's advance warning of a tsunami so move to higher ground.

Feeling shaky...? - If you feel an earthquake in the Pacific Coast area (from Alaska down to Baja), listen to the radio for tsunami warnings.

Is that it...? - Don't be fooled by the size of one wave - more will follow and they could get bigger ... and a small tsunami at one beach can be a giant wave a few miles away!

Be ready to evacuate - Listen to local authorities and leave if you are told to evacuate. *(see EVACUATION)*


## DURING A TSUNAMI:

Leave - If you are told to evacuate, DO IT! Remember - a tsunami is a series of waves - the first one may be small but who knows what the rest will bring! Grab your **Disaster Supplies Kit** and GO!

IF ON OR NEAR SHORE - Get off the shore and get to higher ground quickly! Stay away from rivers and streams that lead to the ocean since tsunamis can go up them too. You cannot outrun a tsunami ... if you see the wave it's too late!

IF ON A BOAT - It depends where you are ... either get to land or go further out to sea!

- In port - You may not have time to get out of port or harbor and out to sea - check with authorities to see what you should do (smaller boats may want to dock and get to land quickly).

- In open ocean - DO NOT return to port if a tsunami warning has been issued since the wave action is barely noticeable in the open ocean! Stay out in open sea or ocean until authorities advise danger has passed.

<u>Don't go there</u> - Don't try to go down to the shoreline to watch and don't be fooled by size of one wave - more will follow and they could get bigger so continue listening to radio and TV.

## AFTER A TSUNAMI:

<u>Listen</u> - Whether on land or at sea, local authorities will advise when it is safe to return to the area -- keep listening to radio and TV updates.

<u>Watch out</u> - Look for downed power lines, flooded areas and other damage caused by the waves.

<u>Don't go in there</u> - Try to stay out of buildings or homes that are damaged until it is safe to enter and wear sturdy work boots and gloves when working in the rubble.

<u>Strange critters</u> – Be aware that the waves may bring in many critters from the ocean (marine life) so watch out for pinchers and stingers!

<u>RED or GREEN sign in window</u> – After a disaster, Volunteers and Emergency Service personnel may go door-to-door to check on people.  By placing a sign in your window that faces the street near the door, you can let them know if you need them to STOP HERE or MOVE ON.

Either use a piece of RED or GREEN construction paper or draw a <u>big</u> RED or GREEN "X" (using a crayon or marker) on a piece of paper and tape it in the window.
- RED means STOP HERE!
- GREEN means EVERYTHING IS OKAY…MOVE ON!
- Nothing in the window would also mean STOP HERE!

<u>Recovery tips</u> - Review TIPS ON RECOVERING FROM A DISASTER at end of this Section.

# What are <u>YOU</u> gonna do about…
## A VOLCANIC ERUPTION?

A volcano is a mountain that opens downward to a reservoir of molten rock (like a huge pool of melted rocks) below the earth's surface. Unlike mountains, which are pushed up from the earth's crust, volcanoes are formed by their buildup of lava, ash flows, and airborne ash and dust. When pressure from gases and the molten rock becomes strong enough to cause an explosion, it erupts and starts to spew gases and rocks through the opening.

Volcanic eruptions can hurl hot rocks (sometimes called **tephra**) for at least 20 miles (32 km) and cause sideways blasts, lava flows, hot ash flows, avalanches, landslides and mudflows (also called **lahars**). They can also cause earthquakes, thunderstorms, flash floods, wildfires, and tsunamis. Sometimes volcanic eruptions can drive people from their homes forever.

Fresh volcanic ash is not like soft ash in a fireplace. Volcanic ash is made of crushed or powdery rocks, crystals from different types of minerals, and glass fragments that are extremely small like dust. But it is hard, gritty, smelly, sometimes corrosive or acidic (means it can wear away or burn things) and does not dissolve in water.

The ash is hot near the volcano but is cool when it falls over great distances. Ashfall is very irritating to skin and eyes and the combination of ash and burning gas can cause lung irritation or damage to small infants, the elderly or people with breathing problems.

**Did you know…**

>    … more than 80 percent of the Earth's surface above and below sea level was formed by volcanic eruptions?!

>    … there are more than 850 active volcanoes around the world and more than two-thirds of them are part of the "Ring of Fire" (a region that encircles the Pacific Ocean)?!

>    … volcanic eruptions can impact our global climate since they release gases like sulfur and carbon dioxide into the earth's atmosphere?!

>    … the primary danger zone around a volcano covers about a 20-mile (32 km) radius?!

>    … floods, airborne ash or dangerous fumes can spread 100 miles (160 km) or more?!

>    … a **pyroclastic** flow is an avalanche of ground-hugging hot rock, ash and gas that races down the slope of a volcano at speeds of 60 mph (97 km/h) with temperatures of nearly 1,300 degrees Fahrenheit (704 degrees Celsius)?!

>    … Alaska has had over 40 active volcanoes?!

# Before a Volcanic Eruption:

Prepare - Review all MITIGATION tips at beginning of this Section. Also try to cover and protect machinery, electronic devices, downspouts, etc. from ashfall. Learn more by visiting the USGS Cascades Volcano Observatory Hazards Safety page at  http://vulcan.wr.usgs.gov/Hazards/Safety

Learn alert levels - Ask local emergency management office which volcano warnings or alert levels are used locally since they vary depending on where you live (can be alert levels, status levels, condition levels or color codes).

Make a plan - Review Section 1 to develop a **Family Emergency Plan** and **Disaster Supplies Kit**. (Note: Put in goggles or safety glasses and dust masks for each family member to protect eyes and lungs from ash.)

Okay to go? - Don't go to an active volcano site unless officials say it's okay.

Be ready to evacuate - Listen to local authorities and leave if you are told to evacuate. *(see EVACUATION)*


# During a Volcanic Eruption:

Listen - Do what local authorities say, especially if they tell you to leave!

Leave - If you are told to evacuate, DO IT! Don't think you are safe to stay at home and watch the eruption… the blast can go for many, many miles and can cause wildfires and many other hazards!

Watch out - Eruptions cause many other disasters:
- **flying rocks** - hurled for miles at extremely fast speeds!
- **mudflows, landslides or lahars** - they move faster than you can walk or run
- **lava flows** - burning liquid rock and nothing can stop it
- **gases and ash** - try to stay upwind since winds will carry these -- they are very harmful to your lungs
- **fires** - hot rocks and hot lava will cause buildings and forests to burn

IF INDOORS - Stay in, but be aware of ash, rocks, mudflows or lava!
- Close all windows, doors, vents and dampers and turn off A/C and fans to keep ash fall out.
- Put damp towels under doorways and drafty windows.
- Bring pets inside (if time - move livestock into closed shelters).
- Listen for creaking on your rooftop (in case ashfall gets heavy -- could cause it to collapse!)

**IF OUTDOORS** - Try to get indoors, if not…

- Stay upwind so ash and gases are blown away from you.
- Watch for falling rocks and, if you get caught in a rockfall, roll into a ball to protect your head!
- Get to higher ground - avoid low-lying areas since poisonous gases collect there and flash floods could happen.
- Use dust-mask or damp cloth over face to help breathing, wear long-sleeved shirts and pants, and use goggles or safety glasses to protect your eyes.
- Ashfall can block out sunlight and may cause lightning.

**IF IN A VEHICLE** - Avoid driving unless absolutely required.

- Slow down and keep speed at 35 mph (56 km/h) or slower, mainly because of thick dust and low visibility.
- Shut off engine and park in garage, if possible (driving stirs up ash that can clog motor and damage moving engine parts).
- Look upstream before crossing a bridge in case a mudflow or landslide is coming.

## AFTER A VOLCANIC ERUPTION:

Listen - Local authorities will say when it's safe to return to area (especially if you had to evacuate) and give other updates when available.

Water - Check with local authorities before using water, even if eruption was just ash fall (gases and ash can contaminate water reserves). Don't wash ash into drainpipes, sewers or storm drains since wet ash can wear away metal.

What to wear - If you must be around ash fall, you should wear long sleeve shirts, pants, sturdy boots or shoes, gloves and keep your mouth and nose covered with a dust-mask or damp cloth.

Ash - Dampen ash before sweeping or shoveling buildup so it's easier to remove and won't fly back up in the air as much - but be careful since wet ash is slippery. Wear protective clothing and dust mask too.

Protect - Cover and protect machinery and electronic devices like computers.

Dust city - Realize ash can disrupt lives of people and critters for months.

Recovery tips - Review TIPS ON RECOVERING FROM A DISASTER at end of this Section. Also visit the USGS Cascades Volcano Observatory Hazards Safety page at  http://vulcan.wr.usgs.gov/Hazards/Safety

# What are <u>YOU</u> gonna do about...
## WINTER STORMS & EXTREME COLD?

Winter storms can last for many days and include high winds, freezing rain, sleet or hail, heavy snowfall and extreme cold. These types of winter storms can shut down a city or area mainly due to blocked roads and downed power lines. People can be stranded in their car or trapped at home for hours or days, but there are many other hazards that come with these storms.

The leading cause of death during winter storms is automobile or other transportation accidents and the second leading cause of death is heart attacks. Hypothermia (or freezing to death) is very common with the elderly who sometimes die inside their homes because it is so cold.

The best way to protect yourself from a winter disaster is to plan ahead before the cold weather begins. Take advantage of spring sales when winter items are cheaper so you are ready for next winter!

## BEFORE A WINTER STORM:

<u>Prepare</u> - Review WIND, FLOOD, and WINTER STORM MITIGATION at beginning of this Section.

<u>Learn the buzzwords</u> - Learn terms / words used with winter conditions...
- **Freezing rain** - rain that freezes when it hits the ground, creating a coating of ice on roads and walkways
- **Hail** - rain that turns to ice while suspended and tossed in the air from violent updrafts in a thunderstorm
- **Sleet** - rain that turns to ice pellets before reaching ground (which can cause roads to freeze and become slippery)
- **Winter Weather Advisory** - cold, ice and snow expected
- **Winter Storm Watch** - severe winter weather such as heavy snow or ice is possible within a day or two
- **Winter Storm Warning** - severe winter conditions have begun or are about to begin
- **Blizzard Warning** - heavy snow and strong winds producing blinding snow (near zero visibility) and life-threatening wind chills for 3 hours or longer
- **Frost/Freeze Warning** - below freezing temperatures expected

<u>Be prepared</u> - Review Section 1 to develop a **Family Emergency Plan** and **Disaster Supplies Kit**, and <u>add</u> the following at home for winter storms:
- **rock salt** - good for melting ice on walkways

- **sand or kitty litter** - to improve traction
- **emergency heating equipment and fuel** - good to have backup in case power is cut off

  <u>fireplace</u> - gas or wood burning stove or fireplace
  <u>generator</u> – gas or diesel models available
  <u>kerosene heaters</u> – ask your Fire Department if they are legal in your community and ask about safety tips in storing fuel!
  <u>charcoal</u> - NEVER use charcoal indoors since fumes are deadly in contained room -- fine for outdoor use!!
- **extra wood** - keep a good supply in a dry area
- **extra blankets** – either regular blankets or emergency blankets (about the size of a wallet)

## DURING A WINTER STORM:

<u>Listen</u> - Get updates from radio and TV weather reports.

<u>What to wear</u> - Dress for the season…
- **layer** - much better to wear several layers of loose-fitting, light-weight, warm clothing than one layer of heavy clothing (outside garment should be waterproof)
- **mittens** - mittens are warmer than gloves
- **hat** - most body heat is lost through the top of your head
- **scarf** - cover your mouth with a scarf or wrap to protect your lungs from cold air

<u>Don't overdo it</u> - Be careful when shoveling snow or working outside since cold can put added strain on heart and cause a heart attack (even in children!)

<u>Carbon monoxide</u> - Learn how to protect your home from winter heating dangers by visiting the Center for Disease Control's Carbon Monoxide web site at  <u>www.cdc.gov/nceh/airpollution/carbonmonoxide</u>

<u>Watch for signs</u> - playing or working out in the snow can cause exposure so look for signs of…
- **frostbite** - loss of feeling in your fingers, toes, nose or ear lobes or they turn really pale
- **hypothermia** - start shivering a lot, slow speech, stumbling, or feel very tired

If signs of either one … get inside quickly and get medical help *(see COLD-RELATED ILLNESSES in Section 3)*

# WINTER DRIVING TIPS

Driving - If you must travel, consider public transportation. Best to travel during the day, don't travel alone, and tell someone where you're going. Stay on main roads and avoid taking back roads.

Winterize car - Make sure you have plenty of antifreeze and snow tires (or chains or cables). Keep gas tank as full as possible during cold weather.

Winter Kit - Carry a "winter" car kit in trunk *(see CAR KIT in Section 1)* and also throw in…

- **warm things** – mittens, hat, emergency blanket, sweater, waterproof jacket or coat
- **cold weather items** - windshield scraper, road salt, sand
- **emergency items** - brightly colored cloth or distress flag, booster cables, emergency flares, tow chain or rope, shovel
- **miscellaneous** - (food, water, etc. mentioned in CAR KIT)

Stranded - If you get trapped in your car by a blizzard or break down…

- **get off the road** - if you can drive, pull the car off main road onto shoulder
- **give a sign** - turn on hazard lights and tie a bright cloth or distress flag on antenna, door handle or hang out driver side window (keep above snow so it draws attention)
- **stay in car** - stay inside until help arrives (your CAR KIT will provide food, water and comforts if you planned ahead)
- **start your car** - turn on car's engine and heater for about 10 minutes each hour (open window slightly for ventilation so you don't get carbon monoxide poisoning)
- **light at night** - turn on inside light at night so crews or rescuers can see you
- **if you walk** - if you walk away from car, make sure you can see building or shelter (no more than 100 yards or 10 m)
- **exercise** - DO NOT overdo it, but light exercises can help keep you warm
- **sleeping** - if others in car, take turns sleeping so someone can watch for rescue crews
- **exhaust pipe** - check exhaust pipe now and then and clear out any snow buildup

## AFTER A WINTER STORM:

Restock - After the storm clears, stock up on items you used so you're ready for the next one!

# Tips on
# Recovering From
# a Disaster...

# Tips on Recovering From A Disaster

Unless you've been in a disaster before, it is hard to imagine how you will handle the situation. Coping with the human suffering and confusion of a disaster requires a certain inner strength. Disasters can cause you to lose a loved one, neighbor or friend or cause you to lose your home, property and personal items. The emotional effects of loss and disruption can show up right away or may appear weeks or months later.

We are going to briefly cover "emotional" recovery tips then cover some "general" recovery tips on what to do AFTER a disaster. Remember -- people *can* and *do* recover from all types of disasters, even the most extreme ones, and you <u>can</u> return to a normal life.

## Emotional Recovery Tips – Handling Emotions

Since disasters usually happen quickly and without warning, they can be very scary for both adults and children. They also may cause you to leave your home and your daily routine and deal with many different emotions, but realize that a lot of this is normal human behavior. It is very important that you understand no matter what the loss is… there is a natural grieving process and every person will handle that process differently.

### Some Normal Reactions to Disasters

<u>Right after disaster</u> – shock, fear, disbelief, has hard time making decisions, refuses to leave home or area, won't find help or help others

<u>Days, weeks or months after disaster</u> – anger or moodiness, depression, loss of weight or change in appetite, nightmares, crying for "no reason", isolation, guilt, anxiety, domestic violence

<u>Additional reactions by children</u> - thumb sucking, bed-wetting, clinging to parent(s) or guardian, won't go to bed or school, tantrums (crying or screaming), problems at school

*Please note:  If any of your disaster reactions seem to last for quite some time, please seek professional counseling to help deal with the problem. There is nothing wrong with asking for help in recovering emotionally!*

### Tips for Adults & Kids

<u>Deal with it</u> - Recognize your own feelings so you can deal with them properly and responsibly.

<u>TALK</u> - Talking to others helps relieve stress and helps you realize you are not alone… other victims are struggling with the same emotions… including

your own family! And don't leave out the little ones … let them talk about their feelings and share your feelings with them.

Accept help - Realize that the people who are trying to help you want to help you so please don't shut them out or turn them away!

Time out - Whenever possible, take some time off and do something you enjoy to help relieve stress… and do something fun with the whole family like a hike, a picnic, or play a game.

Rest - Listen to your body and get as much rest as possible. Stress can run you down so take care of yourself and your family members.

Slow down - Don't feel like you have to do everything at once and pace yourself with a realistic schedule.

Stay healthy - Make sure everyone cleans up with soap and clean water after working in debris. Also, drink lots of clean water and eat healthy meals to keep up your strength. If you packed vitamins and herbs in your **Disaster Supplies Kit**, take them.

Work out - Physical activity - like running or walking - is good for releasing stress or pent-up energy.

Hug - A hug or a gentle touch (holding a hand or an arm) is very helpful during stressful times.

They're watching you - Kids look to adults during a disaster so your reactions will impact the kids (meaning if you act alarmed or worried – they'll be scared, if you cry – they cry, etc.)

Stick together - Keep the family together as much as possible and include kids in discussions and decisions whenever possible.

Draw a picture - Ask your kids to draw a picture of the disaster to help you understand how he or she views what happened.

Explain - Calmly tell your family what you know about the disaster using facts and words they can understand and tell everyone what will happen next so they know what to expect.

Reassurance - Let your kids and family know that they are safe and repeat this as often as necessary to help them regain their confidence.

Praise - Recognizing good behavior and praise for doing certain things (even the littlest of things) will help boost morale.

Watch your temper - Stress will make tempers rise but don't take out your anger on others, especially kids. Be patient and control your emotions.

<u>Let kids help</u> - Including children in small chores during recovery and clean up processes will help them feel like they are part of the team and give them more confidence.

<u>Let others know</u> - Work with your kids' teachers, day-care staff, babysitters and others who may not understand how the disaster has affected them.

## GENERAL RECOVERY TIPS - AFTER A DISASTER

### RETURNING TO A DAMAGED HOME:

<u>Listen</u> - Keep a battery-operated radio with you for any emergency updates.

<u>What to wear</u> – Use sturdy work boots and gloves.

<u>Check outside first</u> - Before you go inside, walk around outside to check for loose power lines, gas leaks, and structural damage.

<u>Call a professional</u> - If you have any doubts about the safety of your home, contact a professional inspector.

<u>Don't go in there</u> - If your home was damaged by fire, do NOT enter until authorities say it is safe (also don't enter home if flood waters remain around the building).

<u>Use a flashlight</u> - There may be gas or other flammable materials in the area so use a battery-operated flashlight (do not use oil, gas lanterns, candles or torches and don't smoke!)

<u>Watch out</u> - Look for critters, especially snakes (flooding will carry them) and use a stick to poke through debris.

<u>Things to check</u> - Some things you want to do first…
- Check for cracks in the roof, foundation and chimneys.
- Watch out for loose boards and slippery floors.
- Check for gas leaks (either by smell or listen for a hissing or blowing sound) ...
  - Start with the hot water heater.
  - Turn off the main gas valve from outside.
  - Call the gas company.
- Check the electrical system (watch for sparks, broken wires or the smell of hot insulation) ...
  - Turn off electricity at main fuse box or circuit breaker.
  - DO NOT touch the fuse box, circuit breaker or wires if in water or if you're wet!

- Check appliances <u>after</u> turning off electricity at main fuse and, if wet, unplug and let them dry out. Call a professional to check them before using.
- Check the water and sewage system and, if pipes are damaged, turn off main water valve.
- Clean up any spilled medicines, bleaches, gasoline, etc.
- Open cabinets carefully since things may fall out.
- Look for valuable items (jewelry, etc.) and protect them.
- Check house for mold. *(see AIR QUALITY MITIGATION)*
- Try to patch up holes, windows and doors to protect home from further damage.
- Clean and <u>disinfect</u> everything that got wet (bleach is best) since mud left behind by floodwaters can contain sewage and chemicals.
- If basement is flooded, pump it out slowly (about 1/3 of the water per day) to avoid damage since the walls may collapse if surrounding ground is still waterlogged. *(see page 62)*
- Check with local authorities about water since it could be contaminated! Wells should be pumped out and the water tested before using, too.
- Throw out food, makeup and medicines that may have been exposed to flood waters and check refrigerated foods to see if they are spoiled. If frozen foods have ice crystals in them then okay to refreeze.
- Call your insurance agent, take pictures of damage, and keep ALL receipts on cleaning and repairs.

## GETTING HELP: DISASTER ASSISTANCE

<u>Listen</u> - Local TV and radio will announce where to get emergency housing, food, first aid, clothing and financial assistance after a disaster.

<u>Help finding family</u> - The Red Cross maintains a database to help people find family, but <u>please</u> don't call office in disaster area since they'll be swamped!

<u>Agencies that help</u> - The Red Cross is often stationed right at the scene of a disaster to help people with immediate medical, food, and housing needs. Some other sources of help include the Salvation Army, church groups and synagogues, and various other Social Service agencies from local, state and provincial governments.

<u>President declares a "Major Disaster" (in U.S.)</u> - In severe U.S. disasters, the government (FEMA) steps in and provides people with ...
- Temporary housing
- Counseling

- Low interest loans and grants
- Businesses and farms are eligible for aid through FEMA

FEMA's Disaster Recovery Centers - FEMA will set up DRCs at local schools and municipal buildings to manually process applications and where people can meet face-to-face with agencies to ...

- Discuss their disaster-related needs.
- Get information about disaster assistance programs.
- Teleregister for assistance.
- Learn about measures for rebuilding that can eliminate or reduce risk of future loss (mitigation tips).
- Learn how to complete Small Business Administration (SBA) loan application (same form used to qualify all individuals for low cost loans or grants, including repair or replacement of damaged homes & furnishings).
- Request status of their Disaster Housing Application.

Or ... people can apply for assistance with DRC over the phone by calling 1-800-621-FEMA (3362).

I lost my job (in U.S.) - People who lose their job due to a disaster may apply for weekly benefits using Disaster Unemployment Assistance (DUA). You can call 1-800-621-FEMA (TTY: 1-800-462-7585) or your local unemployment office for registration information.

Legal help (in U.S.) - Local members of the American Bar Association Young Lawyers Division offer free legal counseling to low-income individuals after the President declares a major disaster. FEMA can provide more information at their DRCs or call 1-800-621-FEMA (TTY: 1-800-462-7585).

Canadian disaster - In the event of a large-scale disaster in Canada, the provincial or territorial government would pay out money to individuals and communities in accordance with its provincial disaster assistance program. *(Federal assistance - Disaster Financial Assistance Arrangements [DFAA] is paid to the province or territory... not to individuals and communities as FEMA does in the U.S.!)*

Recovering financially - The American Red Cross and FEMA developed the following list to help you minimize the financial impact of a disaster:

- **First things first** - 1) remove valuables only if your residence is safe to enter, 2) try to make temporary repairs to limit further damage, and 3) notify your insurance company immediately!
- **Conduct inventory** - make sure you get paid for what you lost
- **Reconstruct lost records** - use catalogs, want ads, Blue Books, court records, request old tax forms from IRS, escrow papers, etc. to help determine value of lost possessions

- **Notify creditors and employers** - let people you do business with know what has happened

- **File insurance claim** - get all policy numbers; find out how they are processing claims; identify your property with a sign; file claims promptly, work with adjusters, etc.

- **Obtain loans and grants** - local media will announce options available for emergency financial assistance

- **Avoid contractor rip-offs** - get several estimates; don't rush into anything; ask for proof of licenses, permits and insurance; get contract in writing; never prepay; get signed release of lien; check out contractors with local Better Business Bureau, etc.

- **Reduce your tax bite** - you may be eligible for tax refunds or deductions but know they can be very complex so you may want to ask an expert for advice

*\* Note: A detailed brochure called "Recovering Financially After a Disaster" prepared by the National Endowment for Financial Education®, the Red Cross, and FEMA is available on the Red Cross's web site or may be at your local Red Cross chapter. (see Section 4)*

## MITIGATION (REDUCING THE IMPACT FOR THE NEXT TIME)

The last thing you want to think about after a disaster is "what if it happens again"! Before you spend a lot of time and money repairing your home after a disaster, find ways to avoid or reduce the impact of the next disaster.

FEMA recommends the following mitigation tips AFTER A DISASTER:

- Ask local building department about agencies that purchase property in areas that have been flooded. You may be able to sell your property to a government agency and move to another location.

- Determine how to rebuild your home to handle the shaking of an earthquake or high winds. Ask local government, hardware dealer, or private home inspector for technical advice.

- Consider options for flood-proofing your home. Determine if your home can be elevated to avoid future flood damage.

- Make sure all construction complies with local building codes that pertain to seismic, flood, fire and wind hazards. Make sure roof is firmly secured to the main frame of the house. Make sure contractors know and follow the code and construction is inspected by a local building inspector.

And please review **ALL** Mitigation tips at the beginning of this Section.

# TIPS ON SHELTER LIVING
## DURING OR AFTER AN EMERGENCY

Taking shelter during a disaster could mean you have to be somewhere for several hours or possibly several days or weeks!  It could be as simple as going to a basement during a tornado warning or staying home without electricity or water for several days during a major storm.

In many emergencies, the Red Cross and other organizations set up public shelters in schools, city or county buildings and churches.  While they often provide water, food, medicine, and basic sanitary facilities, you should plan to have your own supplies - especially water.

Whether your shelter is at home or in a mass care facility use the following tips while staying there during or after an emergency:

Don't leave - Stay in your shelter until local authorities say it's okay to leave. Realize that your stay in your shelter can range from a few hours to weeks ... or longer in some cases!

Take it outside - Restrict smoking to well-ventilated areas (outside if it's safe to go out) and make sure smoking materials are disposed of safely!

Behave - Living with many people in a confined space can be difficult and unpleasant but you must cooperate with shelter managers and others in the shelter.

24-hour watch - Take turns listening to radio updates and keep a 24-hour communications and safety watch going.

Toilet - Bathrooms may not be available so make sure you have a plan for human waste. *(see TIPS ON SANITATION OF HUMAN WASTE)*

Pets - Public shelters do not allow pets due to health reasons (unless it is a service animal assisting a disabled person) so you will have to make arrangements to keep them somewhere else. You can try the Humane Society or local Animal Shelter - if they are still functioning after a disaster.

Next we're going to cover some basic things to think about in the event you and your family are without power, running water, and/or functioning toilets during an emergency or disaster.  We suggest you read over these topics and think about the things you might want to get in advance so you can be prepared for several days or longer.

# Tips on Using Household Foods

## Cooking in a Disaster Situation

When disaster strikes, you may not have electricity or gas for cooking. For emergency cooking you can use a charcoal grill, hibachi or propane camping unit or stove - but only do this <u>OUTDOORS</u>!

**Never** use charcoal in an enclosed environment since it causes deadly fumes!

You can also heat food with candle warmers or a can of sterno.

Canned food can be heated in the can, but remember to remove the paper label and open the can first. And be careful -- don't burn your hand since it may be hot!

Try to limit using salty foods since they can make you thirsty. Also, keep in mind dried foods (like pasta, beans, etc.) require extra water and cooking time so may not be good choices during a disaster situation.

## If the electricity goes off, use Food wisely ...

First - Use perishable food and foods from the refrigerator … and limit opening the frig (don't stand and stare in it like most of us do!) The frig will keep foods cool for about 4 hours without power if left unopened. Dry ice or a block of ice can be placed in refrigerator if power is out more than 4 hours.

Second - Use foods from the freezer and, if possible, have a list of items in the freezer taped outside or at least know how things are organized inside to find stuff quickly. (Keeping door shut keeps cold in!) Foods in a well-filled, well-insulated freezer will not go bad until several days after power goes off. Usually there will be ice crystals in the center of food (which means it's still okay to eat or refreeze) for 2-3 days after a power failure.

Third - Use non-perishable foods and staples in your pantry and cabinets.

## TIP FOR YOUR FREEZER:

Before a disaster strikes, line your indoor and/or outdoor freezer wall with jugs of bottled water. The frozen bottles can help keep food cold longer if you lose power, plus you'll have extra water once it melts. This also helps keep the freezer as full as possible which makes it more energy efficient.

# TIPS ON WATER PURIFICATION

Water is critical for survival. We can go days, even weeks, without food but we <u>must</u> have water to live. For example, an average man (154 pounds) can lose about 3 quarts/litres of water per day and an average woman (140 pounds) can lose over 2 litres - and this could increase depending on your weight and size, on the season, and the altitude.

Your body can lose precious water by sweating and breathing - whether you feel it or not – and, of course, by peeing. In fact, the color of your pee will tell you if you are getting dehydrated. When you drink enough water, your pee will be a light-colored or bright yellow, but when you are dehydrated it will be dark-colored and you'll pee in smaller amounts.

The average person should drink between 2 and 2 ½ quarts/litres of water per day. We suggest you plan on storing about one gallon (4 litres) per day per person to cover for drinking, cooking and personal hygiene - and don't forget water for your pets!

**Did you know…**

> … some 6,000 children die every day from water-related disease?
>
> … about 1.1 <u>billion</u> people (one-sixth of the world's population) don't have access to safe water?

## Use any of the following methods to purify drinking water:

<u>Boiling</u> - Boil vigorously for 5-10 minutes. Boiling water kills most harmful bacteria and parasites. To improve the taste of boiled water pour it back and forth between two containers to add oxygen back into it.

<u>Bleach</u> - Add 10-20 drops of "regular" household bleach per gallon (about 4 litres) of water, mix well, and let stand for 30 minutes. A slight smell or taste of chlorine indicates water is good to drink. *(NOTE: Do <u>NOT</u> use scented bleaches, colorsafe bleaches, or bleaches with added cleaners!)*

<u>Tablets</u> - Use commercial purification tablets and follow instructions. Tablets are pretty inexpensive and found at most sporting goods stores and some drugstores. *(Note: Look for products that contain 5.25 to 6.0 percent sodium hypochlorite as the only active ingredient.)*

<u>Distillation</u> - Involves boiling water and collecting the vapor to remove impurities. *(Check with local library to learn about distillation or visit <u>www.redcross.org</u> and do a SEARCH on **distill**.)*

Also, learn how to remove the water in the hot water heater and other water supplies in your home or office (like ice cubes or the toilet tank - <u>not</u> the toilet bowl and <u>don't</u> use it if chemicals are in the tank!)

# Tips on Sanitation of Human Waste

In disaster situations, plumbing may not be usable, due to broken sewer lines, broken water lines, flooding, or freezing of the system. To avoid the spread of disease, it is critical that human waste be handled in a sanitary manner!

**Did you know...**

... one gram (0.035 oz) of human poop can contain 10 million viruses, 1 million bacteria, 1,000 parasite cysts, and 100 parasite eggs!?[6]

## IF TOILET OKAY BUT LINES ARE NOT...

If water or sewer lines are damaged but the toilet is still intact, you should line the toilet bowl with a plastic bag to collect waste... but DO NOT flush the toilet!! After use, add a small amount of <u>disinfectant</u> to the bag, remove and seal bag (with a twist tie if reusing), and place bag in a tightly covered container away from people to reduce smell.

## IF TOILET IS UNUSABLE...

If toilet is destroyed, a plastic bag in a bucket may be substituted. (Some companies make plastic buckets with snap-on lids.) After use, add a small amount of <u>disinfectant</u> to the bag, and seal or cover bucket.

<u>DISINFECTANTS</u> - **easy and effective for home use in Sanitation of Human Waste. Choose one to store with your Disaster Supplies Kit:**

**Chlorine Bleach** - If water is available, a solution of 1 part liquid household bleach to 10 parts water is best. DO NOT use dry bleach, which is caustic (can burn you, corrode or dissolve things) - is <u>not</u> safe for this kind of use.

**Calcium hypochlorite** - (e.g. HTH, etc.) Available in swimming pool supply or hardware stores and several large discount stores. It can be used in solution by mixing, then storing. Follow directions on the package.

**Portable toilet chemicals** - These come in both liquid and dry formulas and are available at recreational vehicle (RV) supply stores. Use according to package directions. These chemicals are designed especially for toilets that are not connected to sewer lines.

**Powdered, chlorinated lime** - Available at some building supply stores. It can be used dry and be sure to get chlorinated lime - *not* quick lime!

There are also several types of camping toilets and portable toilets available in camping stores and on the Internet that range from fairly low dollars to hundreds of dollars. Or get a small shovel so you can at least dig a hole or latrine outside like campers do.

# TIPS ON HELPING OTHERS
# IN THEIR TIME OF NEED

A disaster really brings out the generosity of many people who want to help the victims. Unfortunately, sometimes this kindness overwhelms agencies that are trying to coordinate relief efforts so please review the following general guidelines defined by FEMA on helping others after a disaster.

- In addition to the people you care for on a daily basis, consider the needs of your neighbors and people with special needs.

- If you want to volunteer your services after a disaster, listen to local news reports for information about where volunteers are needed. Until volunteers are specifically requested, please stay away from disaster areas.

- If you are needed in a disaster area, bring your own food, water and emergency supplies. This is especially important in cases where a large area has been hit since these items may be in short supply.

- Do not drop off food, clothing or other items to a government agency or disaster relief organization unless a particular item has been requested. They usually don't have the resources to sort through donations and it is very costly to ship these bulk items.

- If you wish, give check or money order to a recognized disaster relief organization like the Red Cross. They can process funds, purchase what is needed and get it to the people who need it most. Your entire donation goes towards the disaster relief since these organizations raise money for overhead expenses through separate fund drives.

- If your company wants to donate emergency supplies, donate a quantity of a given item or class of items (such as nonperishable food) rather than a mix of different items. Also, find out where donation is going, how it's going to get there, who's going to unload it and how it will be distributed. Without good planning, much needed supplies will be left unused.

# TIPS FOR VOLUNTEERS
## AND DISASTER WORKERS

- FEMA offers excellent information for disaster workers and volunteers to help you recover emotionally and physically <u>after</u> <u>helping</u> with a disaster.

- FEMA also coordinates many counseling programs to help you adjust back to a normal life since the emotions and images can take weeks or months - even years - to heal.

- Please DO NOT hold these feelings and emotions inside since they can lead to emotional destruction of you and your loved ones through domestic violence, divorce, isolation, addiction, and/or suicide.

- Take advantage of programs and counselors offered by FEMA following a disaster.  There is absolutely nothing wrong with asking for some help in recovering emotionally.

# Section 3

# Information
# & Tips on
# Basic First Aid

# RED CROSS FIRST AID SERVICES AND PROGRAMS

For nearly a century, the American and Canadian Red Cross have trained people in first aid and CPR, translating the consensus of medical science into practical, easy-to-understand information for the public. These first aid and CPR programs are designed to enhance understanding and increase participants' confidence and skill retention.

## AMERICAN RED CROSS HEALTH & SAFETY SERVICES

Some courses are available in both English and Spanish but please check with your Local Red Cross Chapter about availability of Spanish courses.

### First Aid, CPR & AED Courses
**For Your Community:** Basic Aid Training, Community First Aid and Safety, First Aid for Children Today, First Aid - Responding to Emergencies, Infant & Child CPR, Sport Safety Training, etc.

**For Your Workplace:** Core courses like Standard First Aid, Adult CPR/AED, Preventing Disease Transmission, plus various work-related supplemental modules available

**For Professional Rescuers:** Emergency Response, CPR, AED Training, Oxygen Administration, Preventing Disease Transmission

### Swimming & Lifeguarding
Children & Family, Swimming & Fitness, Lifeguard & Aquatic Safety Training, Lifeguard Management

### HIV/AIDS Education
Basic, African American, Hispanic, Workplace

### Caregiving & Babysitting
Babysitter's Training, Nurse Assistant Training, Family Caregiving

### Programs for Youth
HIV/AIDS Programs for Youth, Aquatic Programs for Children & Families, Baby Sitter's Training Course, Youth Services, Street Smart Game, First Aid Matching, Facing Fear: Helping Young People Deal with Terrorism and Tragic Events

### Living Well/Living Safely
Lifeline: Personal Emergency Response Services plus various Fact sheets

# Canadian Red Cross Services

## First Aid Services
**Children Safety Programs:** People Savers, Babysitter's Course, ChildSafe

**First Aid & CPR Programs:** Vital Link Program (Emergency First Aid Course, Standard First Aid Course, CPR/AED Courses, Leadership Program)

**Advanced First Aid Programs:** First Responder, Automated External Defibrillation, Prevent Disease Transmission, Leadership Programs

## Water Safety Services
AquaTots, AquaQuest, AquaSquirts, AquaAdults, AquaLeader, OnBoard

## RespectED
Youth Connect = Relationship Violence & Child Abuse, Educational Presentation Services for youth and community leaders.

## Homecare Services
HomePartners *(Atlantic provinces only)*, Homemakers *(Ontario only)*


Please contact your local Red Cross office in the U.S. or Canada for more information on these courses and to see which programs are available in your area! *(See pages 188 & 195 for contact information!)*

Or you can access more information on the Internet:


**American Red Cross** http://www.redcross.org
*Click on Health & Safety Services*


**Canadian Red Cross** http://www.redcross.ca


*Note: Courses listed here as shown on both Red Cross sites as of May 2004 and subject to change. Please contact local office regarding availability.*

# What are <u>You</u> gonna do about...
## An Emergency?

Everyone should know what to do in an emergency. You should know who to call and what care to provide. Providing care involves giving first aid until professional medical help arrives.

The Emergency Medical Services (EMS) is a network of police, fire and medical personnel, as well as other community resources. People can help EMS by reporting emergencies and helping out victims until EMS can arrive.

During a major disaster, EMS groups will become swamped so if the public is prepared to handle some types of emergencies then we can help some of the victims until EMS arrives.

Your role in the EMS system includes the following things:

BE AWARE...       Realize this is an emergency situation and you could be putting yourself in danger!

BE PREPARED...   Know how to handle the situation.

HAVE A PLAN!     Check **ABCs...**, call 9-1-1 (or call for an ambulance) and help victim, if possible.

## Tips On The ABCs...
## Airway, Breathing & Circulation

In an emergency, you need to check the victim for **ABCs...**:

Airway.        Open the airway by tilting the head back, gently lifting the jaw up, and leaving mouth open.

Breathing.     Place your ear over victim's mouth and nose. Look at chest, listen, and feel for breathing for 3 to 5 seconds.

Circulation.   Check for a pulse using your <u>fingertips</u> (not your thumb) in the soft spot between throat and the muscle on the side of the neck for 5-10 seconds.

# Tips On Making Your "Emergency Action" Plan

**1. BE AWARE**... Make sure it's <u>safe</u> to approach area and victim.

Use your senses...

<u>Listen</u> for cries for help; screams; moans; explosions; breaking glass; crashing metal; gunshots; high winds; popping, humming or buzzing noises; lots of coughing, etc.

<u>Look</u> for broken glass; open medicine cabinet, container or bottle near victim; smoke; fire; vapors or mist; downed power lines, etc.

<u>Watch</u> for signs like trouble breathing; trouble talking; grabbing at throat or chest; pale or blue color in face, lips or ears; lots of people covering mouth or running away, etc.

<u>Smell</u> smoke or something burning; strong odors or vapors (leave if odor is too strong!), etc.

<u>Feel</u> something burning your eyes, lungs or skin, etc.

**2. BE PREPARED**...    The best thing you can do is **STAY CALM**... and <u>THINK</u> before you act!

Any time there's an emergency or disaster, most people are scared or confused and many don't know what to do. Take a few seconds and breathe in through your nose and out through your mouth to help slow your heartbeat and calm down.  <u>Always</u> ask if you can help... either ask the victim or the people around who may be helping.

**3. <u>HAVE A PLAN</u>!**  Check **ABCs**, call 9-1-1 and help victim, if possible.

... Check victims' **ABCs... Airway, Breathing, & Circulation**

... call 9-1-1, 0 for Operator or local emergency number for an ambulance *(see tips on next page)*

... help the victim, if possible

... and STAY with victim until help arrives.

Before giving first aid, you must have the victim's permission.  Tell them who you are, how much training you've had, and how you plan to help.  Do <u>not</u> give care to someone who refuses it - unless they are unable to respond.

# Tips On Calling For An Ambulance

Whenever there is an emergency, you should use the following tips to help decide if you should call 9-1-1 (or your local emergency number) for an ambulance.

## Call if victim...

> … is trapped
>
> … is not responding or is passed out
>
> … is bleeding badly or bleeding cannot be stopped
>
> … has a cut or wound so bad and deep that you can see bone or muscles
>
> … has a body part missing or is torn away
>
> … has pain below the rib cage that does not go away
>
> … is peeing, pooping or puking blood (called passing blood)
>
> … is breathing weird or having trouble breathing
>
> … seems to have hurt their head, neck or back
>
> … is jerking uncontrollably (called having a seizure)
>
> … has broken bones and cannot be moved carefully
>
> … acts like they had a heart attack (chest pain or pressure)

## When you talk to 9-1-1 or the emergency number...

> … try to stay CALM!
>
> … try to describe what happened and what is wrong with the victim
>
> … give the location of the emergency and phone number of where you are calling from
>
> … follow their instructions in case they tell you what to do for the victim.

# Tips On Reducing the Spread of Germs Or Diseases

Whenever you perform first aid on anyone, there is always a chance of spreading germs or diseases between yourself and the victim. These steps should be followed no matter what kind of first aid is being done -- from very minor scrapes to major emergencies -- to reduce the risk of infection.

**BE AWARE...**

> ... Try to avoid body fluids like blood or urine (pee).
>
> ... Cover any open cuts or wounds you have on your body since they are doorways for germs!

**BE PREPARED...**

> ... Wash your hands with soap <u>and</u> water <u>before</u> and <u>after</u> giving first aid.
>
> ... Have a first aid kit handy, if possible.
>
> ... Put something between yourself and victim's body fluids, if possible ...
>
>> <u>blood or urine</u> - wear disposable gloves or use a clean dry cloth
>>
>> <u>saliva or spittle</u> – use a disposable Face Shield during Rescue Breathing
>
> ... Clean up the area with household bleach to kill germs.

**... and... HAVE A PLAN!**

> ... *see TIPS ON MAKING <u>YOUR</u> "EMERGENCY ACTION" PLAN two pages back.*

# Tips On Good Samaritan Laws

The definition of a "Samaritan" is a charitable or helpful person. Most states have Good Samaritan laws that were designed to protect citizens who try to help injured victims with emergency care.

If a citizen uses "logical" or "rational" actions while making wise or careful decisions during an emergency situation then they can be protected from being sued.

To learn more about your state's Good Samaritan laws, check with your local library or contact an attorney.

# Tips on
# Basic First Aid...

# What are <u>YOU</u> gonna do about…
## BITES & STINGS?

## ANIMAL & HUMAN BITES

Americans and Canadians report approximately 5 million bites each year (mostly from dogs). Both humans and animals carry bacteria and viruses in their mouths, however, human bites are more dangerous and infection-prone because people seem to have more reactions to the human bacteria.

**Things to watch for…**
> **Puncture or bite marks**
> **Bleeding**
> **Infection** - Pain or tenderness, redness, heat, or swelling, pus, red streaks
> **Allergic Reaction** - Feeling ill, dizzy or trouble breathing

**What to do…**
- Wash the bite as soon as possible to remove saliva and dirt from the bite wound  - use running water and soap or rinse area with hydrogen peroxide.
- Control bleeding using direct pressure with cloth or gauze.
- Pat dry and cover with sterile bandage, gauze, or clean cloth - don't put a cream or gel on wound -- may prevent drainage.
- May want to raise bitten area to reduce swelling or use a cold pack or cloth with an ice cube in it.
- Call local emergency number or call your Animal Control* (usually listed in blue Government pages in phone book under County / Municipality).
- Watch for any allergic reactions for few days (see list above).

*…also…*
- Get to a doctor or hospital if bleeding is really bad, if you think animal could have rabies, or if stitches are required!

*\* Note:  For Human bites, there is no need to call Animal Control, however, most states/provinces <u>do</u> require all animal and human bites be reported to local police or health authorities!*

## INSECT BITES & STINGS

First we cover first aid for bites & stings in general, then "West Nile Virus".

**Things to watch for…**
> **Stinger**

**Puncture or bite mark**
**Burning pain or Swelling**
**Allergic Reaction** - Pain, itching, hives, redness or discoloration at site, trouble breathing, signs of shock (pale, cold, drowsy, etc.)

**What to do...**

- Remove stinger by scraping it away with credit card, long fingernail or using tweezers. Don't try to squeeze it out since this causes more venom to get in the victim.
- Wash wound with soap and water or rinse with hydrogen peroxide.
- Cover with sterile bandage or gauze, or clean cloth.
- Place a cold pack on the bandage (a baggie or cloth with ice works fine).
- Watch for any allergic reactions for a few days (see above).

*(See WEST NILE VIRUS on next page too. TICKS are covered on page 146.)*

**To relieve pain from an insect bite or sting:**

**Activated charcoal** - Empty 2-3 capsules into a container and add a small amount of warm water to make a paste. Dab paste on the sting site and cover with gauze or plastic to keep it moist. (Note: powder makes a black mess but it's easily wiped off with a towel!) This will help draw out the venom so it collects on your skin. *(See FIRST AID KITS in Section 1)*

**Baking Soda** - Make a paste of 3 parts baking soda + 1 part warm water and apply to the sting site for 15-20 minutes.

**Clay mudpack** - If in the wilderness, put a mudpack over injury and cover with bandage or cloth. The mudpack must be a mix of clay-containing soil since clay is the key element but don't use if any skin is cracked or broken.

**Meat tenderizer** - Mixing meat tenderizer (check ingredient list for "papain") with warm water and applying to the sting will help break down insect venom. (Papain is a natural enzyme derived from papaya.)

**Urine (Pee)** - Another remedy useful in the wilderness which sounds totally gross... but has a history of medical applications in a number of cultures... is urine which reduces the stinging pain. (Unless you have a urine infection the urine will be sterile and at the least won't do any harm!)

**Some potential pain-relieving and anti-inflammatory remedies include:**

**fresh aloe** - break open a leaf or use 96-100% pure aloe gel
**lemon juice** - from a fresh lemon

**vitamin C** - make paste with 3 crushed tablets + drops of warm water

**vitamin E** - oil from a bottle or break open a few gel capsules

**store brands** - if over-the-counter methods preferred, use calamine cream or lotion and aspirin or acetaminophen


## WEST NILE VIRUS

The CDC reports 9,006 cases of West Nile in 2003 (double 2002 figures) with 220 deaths and 2,695 cases of severe brain damage. Health Canada reports over 1,300 cases across 7 provinces with no known deaths.

West Nile virus is primarily spread by mosquitoes that have fed on infected birds. But realize, out of the 700+ species of mosquitoes in the U.S. (and the 74 species in Canada), very few - less than 1% - will become infected with the virus. And out of all the people bitten by an infected mosquito, less than 1% will become severely ill so chances are pretty slim you would ever get it.

The virus usually causes fever, aches and general discomfort. Severe cases can cause inflammation of the lining of the brain or spinal cord (meningitis) or inflammation of the brain itself (encephalitis) - either can be fatal. People with weakened immune systems, children and seniors are at greatest risk.

**Things to watch for...**
> *(Most symptoms appear within 3-15 days after being bitten)*
> **Mild flu-like symptoms** - fever, headache and body aches
> **Mild skin rash and swollen lymph glands**
> **Severe symptoms** - severe headache, high fever, neck stiffness, confusion, shakes, coma, convulsions or muscle weakness, meningitis or encephalitis

**What to do...**
- There is no "cure" other than a body fighting off the virus naturally - mainly just watch symptoms.
- Consider boosting immune system to help fight virus (like taking astragalus, Vitamin C, garlic, mushrooms, zinc, a good multiple vitamin + mineral supplement, etc. - but check with doctor if taking prescription medications).
- If **mild** symptoms appear, keep watching person for a few weeks in case symptoms get worse.
- If **severe** symptoms appear, get medical attention quickly since it could become deadly.

**Things to do to avoid mosquito bites ...**
- Stay indoors at dawn, dusk, and early evenings when mosquitoes are most active.
- Wear long-sleeved shirts and long pants when outdoors.

- Spray clothing and exposed skin with repellent containing DEET (N,N-diethyl-meta-toluamide) -- the higher % of DEET, the longer you're protected from bites (20% lasts about 4 hours)
- Don't put repellent on small children's hands since it may irritate their mouths or eyes.
- Get rid of "standing water" sources around yard and home since they are breeding grounds for mosquitoes.
- The CDC says Vitamin B and "ultrasonic" devices are NOT effective in preventing mosquito bites!

There is a growing concern about mosquitoes and other vectors spreading diseases or being used as "bioterrorism", but due to space we are only listing West Nile Virus for now. Always listen to health authorities for information on mosquito-borne and other types of infections.

*For more information about West Nile virus and other diseases visit the CDC NCID's Division of Vector-Borne Infectious Diseases web site at www.cdc.gov/ncidod/dvbid ...or... Health Canada's site at www.hc-sc.gc.ca (click on "Diseases & Conditions")*

## SEA CRITTER (MARINE LIFE) STINGS

There are too many types of sea critters in our oceans and seas so we can't cover all the various types of stings and bites that could happen, but a few common stings are shown below. If you want to learn more about specific types of sea critters (actually called marine life) then check with your local library or on the Internet.

**Things to watch for...**
> **Puncture marks or tentacles on the skin**
> **Pain or burning**
> **Swelling or red marks**
> **Possible Allergic Reaction** - Pain, itching, hives, redness or discoloration at site, trouble breathing, signs of shock (pale, cold, drowsy, etc.)

**What to do...**
- Rinse skin - use seawater, vinegar, ammonia, or alcohol (in whatever form is handy - rubbing alcohol or liquor!) Fresh water might make it hurt worse!
- DO NOT rub skin - it could make it worse!
- Try to remove any tentacles attached to skin, if possible... but DO NOT use bare hands... use a towel or tweezers!
- Soak sting or make a paste (see below) to help relieve pain:
  **tropical jellyfish** - soak area in vinegar
  **stingray or stonefish** - soak area in hot water

- Cover sting with sterile bandage or gauze, or clean cloth.
- Call local emergency number, if necessary.

**To relieve pain from a sea critter sting:**

**Baking Soda Paste** - Make a paste of 3 parts baking soda plus 1 part warm water and apply to sting site until it dries. Scrape off paste with a knife or credit card to help remove some of the skin. (Note: two other quick and easy pastes are sand and seawater or flour and seawater! Scrape off as above.)

**Urine (Pee)** – Again, we know this sounds weird… but urine (pee) has a history of medical applications and can reduce stinging pain. (Unless you have a urine infection, it will be sterile and at the least won't do any harm!)

## SNAKE BITES
According to the FDA, between 7,000 - 8,000 people in the U.S. are treated for poisonous snake bites each year usually resulting in less than 10 deaths. Snake bites can cause infection or allergic reaction in some people -- even bites from non-venomous snakes.

Poisonous snakes have triangular heads, slit-like pupils, and two long fangs which make puncture wounds at end of each row of teeth. Non-poisonous snake bites leave two rows of teeth marks but no puncture wounds, but don't use bite mark to determine type of snake since swelling could hide wounds.

**Things to watch for…**
> **Puncture and/or bite marks**
> **Pain and Swelling**
> **Nausea and puking**
> **Difficulty breathing or swallowing**
> **Possible Allergic reaction** – Weakness or dizziness;
> redness or discoloration at bite; trouble breathing;
> signs of shock (pale, cold, drowsy, etc.)

**What to do…**
- If possible, try to identify type or color of snake but don't put yourself in danger!
- Wash bite wound with soap and water.
- Keep bitten body part (hand, etc.) below heart level, if possible.
- Call local emergency number or animal control, if necessary.

**If bite is from a Poisonous snake, also do this…**
- Remove constrictive items (like rings or watches) since swelling may occur.
- DO NOT apply tourniquet or ice!

- Monitor breathing and make sure airway is open.
- Keep victim as still as possible to slow down the circulation of venom.
- DO NOT let victim eat or drink anything or take medication since it could interfere with emergency treatment!
- If possible and safe, remove venom - esp. if help is hours away! (Most snakebite kits have proper venom extractors in them.)
- DO NOT use "cut and suck" method... may cause infection!
- Get to a doctor or hospital to receive antivenin.

The worst effects may not be felt for hours after a bite from most poisonous North American snakes, but it is best if antivenin is given as quickly as possible (or at least within 12-24 hours of the bite).


## Spider Bites, Scorpion Stings, & Ticks

There are only a dozen or so spiders that can actually cause symptoms or side effects to humans with a bite -- the most serious are black widows and brown recluses. Tarantulas are also a little serious but do not cause extreme reactions and rarely will kill a human.

Scorpions will sting anything that touches them. Their sting feels like a small electrical shock (almost like a hot needle). Scorpions whip their tails over their body and zap their enemy many times, but it happens so quick it may only feel like one sting!

You may think ticks are insects but they're actually bloodsucking arachnids. Adult ticks have eight legs and two body segments just like spiders, mites and chiggers. Ticks grab onto a host (animals or people walking through brush) and sink their harpoon-like barbed mouth and head into the host's skin until full of blood. Then they drop off and wait for their next meal to pass by. Since ticks feast on one spot for days, they can spread bacteria and diseases from host to host (like from animals to humans) - even by touching them.

The main threat of both spiders and scorpions is the allergic reaction humans have to their bite or sting so symptoms need to be watched carefully. Obviously the main threat of ticks is the risk of disease or illness (like lyme disease, Rocky Mountain spotted fever or tick paralysis).

**Things to watch for...**

> **Bite or sting mark or ticks**
> **Pain or burning feeling**
> **Redness or Swelling or Rash**
> **Stomach pain or puking**
> **Flu-like symptoms** - fever, dizziness, weakness, headache,
> body aches, swollen lymph nodes, etc.

**Change in skin color or bruising or rash** (may look kind of like a bulls-eye)

**Possible Allergic reactions** - trouble breathing or swallowing, signs of shock (pale, cold, etc.)

## What to do for SPIDERS and SCORPIONS...

- Try to identify the type of spider or scorpion, but don't put yourself in danger!
- Wash bite wound with soap and water or rubbing alcohol.
- Apply a cold pack (a baggie or cloth with ice will work).
- Get to a doctor or hospital to receive antivenin (if a poisonous spider / scorpion) or call local emergency number, if necessary.
- Watch for allergic reactions or infections for several days.

## What to do for TICKS...

Key things are to find a tick before it feasts for days and to remove a tick slowly with head intact so it doesn't spew bacteria into the blood stream.

- DO NOT use petroleum jelly, nail polish or heat - don't work!
- Use tweezers or commercial tick remover (or cover fingers).
- Grasp tick close to skin where head is buried - don't squeeze it!
- Slowly pull the tick straight up until skin puckers (may take several seconds but tick will loosen it's barbs and let go).
- DO NOT throw tick away - may need it tested! Put in zippered baggie with moist paper towel, date it, and put in refrigerator.
- Wash bite wound and tweezers with soap and water.
- Call local health department or vet to ask if tick needs to be identified or tested. If not, throw away baggie.
- Watch for rash, infection or other symptoms for a week or so.

## Things to do to avoid ticks ...

- Wear light-colored pants & long-sleeve shirt (to see ticks), a hat (to keep out of hair) and tuck in (pants in socks + shirt in pants).
- Do full body checks couple times a day during tick season.
- Use tick repellent with DEET (N,N-diethyl-meta-toluamide).

*See Page 144 for CDC NCID's and Health Canada's web sites to learn more.*

## To relieve pain from spider or tick bite or scorpion sting:

**Ammonia** -- Place a small amount of ammonia on a cotton ball and apply directly on the bite or sting for a few seconds to reduce stinging pain.

**Tea Tree Oil** -- Apply a few drops of 100% Melaleuca alternifolia (Tea Tree oil) directly on bite or sting but avoid getting it in or near your eyes.

# What are <u>YOU</u> gonna do about...
## BLEEDING?

## CONTROLLING BLEEDING

**Things to watch for...**
> **Source of bleeding**
> **Pain and/or Swelling**
> **Object sticking out or stuck in wound** (like a piece of metal or glass or a bullet)
> **Shock** (pale, cold or clammy, drowsy, weak or rapid pulse, etc.)

**What to do...**
- Be aware of your surroundings and be prepared to call an ambulance. (*see TIPS ON CALLING FOR AN AMBULANCE*)

**If there IS object sticking out of wound (or possibly deep inside):**
- Put thick soft pads around object sticking out (or around wound).
- Gently try to apply pressure to help stop the bleeding.
- DO NOT try to remove or press on the object!
- Carefully wrap with a roller bandage to hold the thick pads around the object.
- Get medical attention immediately!

**If there is NO object sticking out of the wound:**
- Be careful since there might be something inside the wound that you can't see!
- Cover wound with a clean cloth or sterile gauze pad and press firmly against the wound... and follow above steps if victim has an object <u>inside</u> the wound.
- If cloth or gauze becomes soaked with blood, DO NOT remove it! Just keep adding new dressings on top of the old ones.
- You want to try and carefully raise or elevate the injured body part above the level of the victim's heart but be aware...there may be broken bones.
- Keep applying pressure on dressings until bleeding stops.
- Use a firm roller bandage to cover gauze or cloth dressings.

**If bleeding won't stop:**
- Put pressure on a nearby artery to help slow down blood flow
  <u>Arm</u> – press inside of upper arm, between shoulder and elbow
  <u>Leg</u> – press area where the leg joins front of the hip (groin)

# INTERNAL BLEEDING

Minor internal bleeding is like a bruise - a vein, artery or capillary can break or rupture spewing blood under the skin. A more serious form of internal bleeding can be caused by a major fall, crushing accident or a blow to the head. It's very hard to tell if a person is suffering from internal bleeding since there may not be blood outside the body. Symptoms don't always appear right away but can be life-threatening so get medical help quickly.

**Things to watch for...**
> **Abdominal pain or tenderness**
> **Pain and/or Swelling in abdomen** (around belly button)
> **Shock** (pale, cold or clammy, drowsy, weak or rapid pulse, etc.)
> **Either a fast or slow pulse**
> **Coughing up bright, foamy blood** (if dark red means bleeding inside for a while)
> **Blood shows up in pee, poop or puke** (in urine, stool or vomit)

**What to do...**
- Be aware of your surroundings and call for an ambulance.
- Don't move victim if injuries to head, neck or spine .
- Check **ABCs... Airway, Breathing** & **Circulation**.
- Stay with victim until help arrives.

*(Please review HEAD, NECK & SPINE INJURIES and SHOCK too)*


# NOSEBLEEDS

**What to do...**
- Pinch the soft part of the nose for about 10 minutes.
- Have the person sit and lean forward.
- Put an icepack or cold compress on the bridge of the nose.


# SLASHED OR SEVERED BODY PARTS/AMPUTATION

**What to do...**
- Keep direct pressure on the stump to stop the bleeding.
- Find the body part, if possible, and wrap it in gauze or a clean cloth.
- Put body part in an airtight plastic bag, put the bag in ice water and take it to the hospital with the victim.

# What are <u>YOU</u> gonna do about...
## BREATHING PROBLEMS?

### ASTHMA ATTACK

**Things to watch for...**
> **Noisy breathing or wheezing**
> **Difficulty in breathing or speaking**
> **Blueness of skin, lips and fingertips or nails**

**What to do...**
- Make sure victim has nothing in mouth (keep an open airway).
- Have victim sit up straight to make breathing easier.
- If victim has medication, or an inhaler, have them take it.
- Try to keep victim and yourself calm!
- If attack is severe, call for an ambulance or emergency help.

**Some tips that could possibly help slow down an asthma attack:**
*(Note: These tips are NOT to be used as a replacement for medical attention but could be helpful in the early stages of an asthma attack.)*

**Pursed lip breathing** - At the first sign of an attack, breathe in deeply through the nose and out through the mouth with lips pursed (like blowing up a balloon). It will help relax the body and possibly get rid of stale air in the lungs.

**Drink a warm liquid or caffeine** - Drinking one or two cups of coffee or tea that have caffeine could help relax the bronchial tubes. If you decide to drink a soda, do not use ice since cold could possibly trigger an attack - the warmer the better.

*For more information about Asthma visit the CDC NCEH's Air Pollution and Respiratory web site at: www.cdc.gov/nceh/airpollution/asthma*

### RESCUE BREATHING (NOT BREATHING)
Rescue breathing (or mouth-to-mouth resuscitation) should only be done when the victim is not breathing on his or her own. Make sure the victim is not choking on anything like vomit, blood or food *(if so, see CHOKING)* and check them using the **ABCs... Airway**, **Breathing**, and **Circulation**!

**Things to watch for...**
> **Grabbing at throat**
> **Can't feel, see or hear any breaths**

**Trouble breathing or talking**
**Bluish color of skin, lips, fingertips or nails, and earlobes**

**What to do…**
- BE AWARE… make sure there is no head or neck injury first!
- Carefully move victim so they are flat on their back.
- Tilt adult's head all the way back and lift chin. (Be careful with a child's or infant's head… just tilt head a little bit!)
- Watch chest, listen, and feel for breathing for about 5 seconds.

**If victim is NOT breathing begin Rescue Breathing…**
- Pinch victim's nose shut.
- Open your mouth wide to make a tight seal around the victim's mouth *(Note: For infant, cover both mouth and nose with your mouth.)*
- Give victim 2 slow breaths to make their chest rise.
- Check for pulse using your fingers in soft spot between the throat and the muscle on side of the neck for 5-10 seconds.
- Continue Rescue Breathing if victim has a pulse but is not breathing…
  Adult - give 1 breath every 5 seconds.
  Child or Infant - give 1 breath every 3 seconds.
- Check pulse and breathing every minute until the victim is breathing on their own.

***…also…***
- If victim pukes… turn them gently on their side, wipe mouth clean, turn them back and continue Rescue Breathing until they are breathing on their own.

**NOTE: If victim NOT breathing and DOES NOT have pulse, see HEART EMERGENCIES for tips on Giving CPR!**

# What are <u>YOU</u> gonna do about...
## BROKEN OR FRACTURED BONES?

A fracture is the same as a break and can range from a small chip to a bone that breaks through the skin. If you suspect a fracture, use a splint to keep the victim from moving too much and get professional help... and let the trained medical experts decide what is wrong!

**Note: Also review HEAD, NECK OR SPINE INJURIES if needed.**

**Things to watch for...**
> **Pain, bruising or swelling**
> **Bleeding**
> **Limb or area moves strange or looks strange**
> **Shock** (pale, cold or clammy, drowsy, weak or rapid pulse, etc.)

**What to do...**
- DO NOT move bone or try to straighten limb if bone breaks through skin!
- Try not to move victim unless they are in danger.
- Have victim sit or lie down to rest the injured part.
- If possible, raise or elevate the injured part.
- Put a cold compress or ice pack on injury to reduce swelling.
- If help is delayed or you need to move victim, splint injury the same way it was found.
- Be prepared to call an ambulance, if necessary.

## TIPS ON SPLINTING

A splint can be made using magazines, newspapers, a pillow, wood, etc.

**Some basic tips on splinting include...**
> ... always splint an injury the same way it was found
> ... make sure item being used for splint is longer than the broken bone
> ... use cloth strips, neck ties, thin rope, etc. for ties
> ... put something soft between the splint and the bone
> ... tie the splint <u>above</u> and <u>below</u> the break... but don't tie it too tight!
> ... touch an area below the splint and ask victim if they can feel it... if not, loosen ties!
> ... put a cold compress or ice pack on the injury and keep victim warm with a blanket or whatever is available.

# What are <u>YOU</u> gonna do about...
## BURNS?

Depending on how bad a burn is will determine what it is called:

**First degree burns** - hurts only top layer of skin; turns pink or red; some pain and swelling; no blisters (usually from sun, chemicals, or touching something hot)

**Second degree burns** - hurts the two upper layers of skin; very painful and causes swelling that lasts several days; blisters and possibly scars (usually from deep sunburn, chemicals, fire or hot liquid spills)

**Third degree burns** - hurts all skin layers and possibly tissue; charred, raw or oozing areas; destroys cells that form new skin; nerve cells are destroyed and can take months to heal (usually from being exposed to fire or electrical shock for a long time). Can cause severe loss of fluids, shock and death.

## BURNS FROM FIRE OR HOT LIQUIDS

**Things to watch for...**
> **Skin is red and swollen**
> **Blisters that may open and ooze clear or yellowish fluid**
> **Minor to Severe Pain**

**What to do...**
- BE AWARE... and don't put yourself in danger!
- Stop the burning by putting out flames and move victim from source of the burn. (If victim is on fire, tell them to STOP, DROP and ROLL!)
- Cool burn by using large amounts of <u>running</u> cool water for about 10 minutes. For hard to reach areas, wet a cloth, towel or sheet and carefully keep adding water!
- Try to remove clothing, rings or jewelry in case of swelling (DO NOT remove any items that are stuck to burned areas!)
- Cover burn with a sterile bandage or clean cloth. (Try to keep fingers and toes separated with bandage or cloth, if possible.)
- Seek medical attention, if necessary.

**Things you should NOT do...**
- DO NOT break any blisters!
- DO NOT remove any item that sticks to the skin!
- DO NOT apply any creams, oils or lotions to the burns - wait for the medical experts!

# CHEMICAL BURNS

*Also see TERRORISM (in Section 2) for information, signs & symptoms, and treatment for several chemical agents in liquid, solid or aerosol forms that may cause chemical burns.*

**Things to watch for...**
> **Rash or blisters**
> **Trouble breathing**
> **Dizziness or headache**
> **Name of the chemical**

**What to do...**
- Rinse area with cool running water for at least 15 minutes.
- Remove any clothing, rings or jewelry that may have the chemical on it.
- Make a note of chemical name for medical staff or hospital.

# ELECTRICAL BURNS

**Things to watch for...**
> **Electrical appliances or wires**
> **Downed power lines**
> **Sparks and/or crackling noises**
> **Victim may have muscle spasms or trembling**
> **Lightning during a storm**

**What to do...**
- BE AWARE... and don't put yourself in danger!  If a power line is down, wait for the Fire Department or Power Company.
- DO NOT go near victim until power is OFF!  Pull plug at wall outlet or shut off breaker. Once off, it's okay to touch victim.
- If victim was struck by lightning, they CAN be touched safely!
- Check **ABCs ... Airway, Breathing,** & **Circulation** if victim is passed out - you may need to do Rescue Breathing or CPR. *(see BREATHING PROBLEMS and HEART PROBLEMS)*
- Don't move victim unless they are in danger.
- There should be 2 wounds - usually have enter and exit burns.
- DO NOT try to cool the burn with anything!
- Cover burn with a dry sterile bandage or clean cloth.
- Seek medical attention, if necessary.

# SUNBURN

Sunblocks and lotions should be applied at least 20 minutes <u>BEFORE</u> going in the sun so it can be absorbed into the skin layers, especially little ones!

Remember… dark colors absorb heat so best to wear light or white colors to reflect sunlight. And you can get sunburned on a cloudy day just as easily as a sunny day - if you can see a shadow, you're still catching some rays.

**Things to watch out for…**
>**Blisters or bubbles on the skin**
>**Swelling or pain**

**What to do…**
- Cool the burn by using cool cloths or pure aloe vera gel.
- Get out of sun or cover up so you won't get further damage.
- Take care of blisters by loosely covering them and don't pick at them!

<u>**To help relieve the pain from a sunburn if NO blisters exist:**</u>

**Aloe vera** - Break open a fresh leaf or use 96-100% pure aloe gel.

**Baking soda** - Add ½ cup baking soda to a warm bath and soak for half an hour.

**Vinegar -** Put some regular or cider vinegar on a cloth and apply to sun-burned area.

**Whole milk** - Apply a cool compress soaked in whole milk to the area.

# What are <u>YOU</u> gonna do about...
## CHOKING?

**Things to watch for...**
>  **Trouble breathing**
>  **Coughing or choking for several minutes**
>  **Gripping the throat with one or both hands**
>  **High-pitched wheezing**
>  **Bluish color of skin, lips, fingertips or nails, and earlobes**

ATTENTION: There are TWO separate "**What to do...**" parts here... one for <u>ADULTS & CHILDREN</u> (below) and one for <u>INFANTS</u> (see next page)!

**What to do... for <u>ADULTS & CHILDREN</u>** *(Children over age 1)*
- Tell victim to try and cough it out.
- If victim stops breathing, then BE PREPARED to give the Heimlich maneuver (next 2 bullets) and tell someone to call an ambulance.
- Stand behind victim and place your fist (thumb side in) just above victim's belly button.
- Grab your fist with your other hand and give quick, upward thrusts into their stomach until object is coughed up (or the victim passes out).

**If the <u>ADULT or CHILD</u> passes out:**
- Check for an object in victim's mouth and try to clear it out with your fingers.
- Begin Rescue Breathing. *(see BREATHING PROBLEMS)*

**If NO air gets in <u>ADULT or CHILD</u> during Rescue Breathing:**
*Combine Heimlich maneuver with Rescue Breathing*
- Put heel of one hand just above victim's belly button and put your other hand on top of the first.
- Give about 6-10 upward thrusts to try to clear their windpipe.
- Check for an object in victim's mouth and try to clear it out with your fingers.
- Try to give Rescue Breathing again to see if air will go in.
- Continue above steps until victim can breathe on their own or until help arrives.

## What to do... for <u>INFANTS</u> *(Newborn to age 1)*

- If infant stops breathing, have someone call for an ambulance.
- Turn infant face down on your forearm and support its head with that hand -- hold at angle so it's head is lower than chest. (May want to brace arm holding infant against your thigh!)
- Give 5 back blows between infants' shoulder blades with heel of your other hand.
- Turn infant over so it is facing up on your forearm (still at an angle so head lower than chest) -- use your **first two fingers** to find center of the breastbone on infant's chest.
- Give 5 thrusts to infant's chest using **only** **2 fingers**! (Each thrust should be about 1/2 to 1 inch [1 to 2.5 cm] deep!)
- Repeat all steps until object is coughed up or infant passes out.

## If the <u>INFANT</u> passes out:

- Check for object in the infant's mouth and try to clear it out with your fingers.
- Begin Rescue Breathing and remember... cover **both** mouth <u>and</u> nose on Infants! *(see BREATHING PROBLEMS)*

## If no air gets in <u>INFANT</u> during Rescue Breathing:

- Turn infant face down on your forearm again supporting its head with that hand -- hold at angle so head lower than chest.
- Give 5 back blows between infants' shoulder blades with heel of your other hand.
- Turn infant over so it is facing up on your forearm and use your **first two fingers** to find center of the breastbone on infant's chest.
- Give 5 thrusts to infant's chest using **only** **2 fingers**!
- Try to give Rescue Breathing again to see if air will go in.
- Continue above steps until infant can breathe on its own or until help arrives.

# What are <u>YOU</u> gonna do about...
## COLD-RELATED ILLNESSES?

### FROSTBITE

Frostbite (or frostnip which is the early stages of frostbite) is when certain parts of your body are exposed to severe or extreme cold - mainly your fingers, toes, ears, cheeks and nose. Freezing temperatures can form ice crystals in the fluids in and around cells in your body. This damages and dries out cell tissues and membranes, and extreme cases can impact deep nerves, muscles or even bones... or even lead to the loss of a limb!

**Things to watch for...**
> **Skin appears white and waxy**
> **Numbness or no feeling in that area**
> **Possible blisters**

**What to do...**
- Handle area gently; DO NOT rub the affected area.
- Remove tight or constrictive clothing (gloves, boots, socks, etc.) and any jewelry.
- Warm gently using body heat or soaking area in warm water (between 100-105 degrees Fahrenheit / between 38-41 degrees Celsius) until area is red and feels warm. *(Person may feel a burning sensation or pain as the area warms back up.)*
- Loosely bandage the area with dry, sterile dressing or cloth.
- If fingers or toes are frostbitten, separate them with sterile gauze or clean cloth.
- Try not to break any blisters.

**Things you should NOT do...**
- DO NOT rub or massage the area since this may cause damage to cells!
- DO NOT rub snow on the area!
- DO NOT try to warm with dry radiant heat (meaning don't warm with a blow-dryer or hold in front of fire or hot stove). Using warm water is best!
- DO NOT try to thaw a frostbitten body part if it has a chance of re-freezing (if you are stuck in the wilderness) since this could cause more damage.

# HYPOTHERMIA

Hypothermia can start setting in when your body core (the vital organs - heart, lungs, and kidneys) drops below 95 degrees Fahrenheit (35 degrees Celsius). When exposed to extreme cold for a long time, your brain begins to shut down certain bodily functions to save internal heat for the body core.

**Things to watch for…**
> **Shivering and numbness**
> **Confusion or dizziness**
> **Stumbling and weakness**
> **Slow or slurred speech**
> **Shock**  (pale, cold or clammy, drowsy, weak or rapid pulse, etc.)

**What to do…**

- Gently move victim to a warm place.
- Check breathing and pulse (**ABCs … Airway, Breathing,** & **Circulation**).
- Handle victim gently and DO NOT rub body or limbs.
- Remove any wet clothing and replace with dry clothing and/or blankets.
- If possible, place victim in a sleeping bag, especially if in the wilderness. (Note: Your body heat can help heat victim… so cuddle up - if victim says it's okay!)
- Cover the head and neck with a hat or part of a blanket (75% of the body's heat is lost through top of the head).
- DO NOT WARM VICTIM TOO QUICKLY, such as putting them in warm water! (If the body warms too fast, it can dump the cold blood into the heart and body core causing a possible heart attack or drop in body temperature.)
- If hot water bottles or hot packs are used, wrap them in a towel or blanket first then place them on side of the chest or on groin (hip) area. (If heat is put on arms or legs then blood could be drawn away from body core - keep heat on the core!)
- Let victim sip a warm, sweet, <u>nonalcoholic</u> drink.
- Keep watching victim's **ABCs...**

**Things you should NOT do…**

- DO NOT rub or massage the victims limbs!
- DO NOT put victim in a hot bath! It will warm him/her TOO quickly!
- DO NOT put hot packs on arms or legs… put them against the body (chest or groin area)!

# What are <u>YOU</u> gonna do about…
## CONVULSIONS & SEIZURES?

## CONVULSIONS

A convulsion is usually brought on by a high fever, poisoning or injury and is basically a violent seizure *(see "Things to watch for…")*.

## SEIZURES

Seizures are usually related to epilepsy (also known as seizure disorder since seizures occur repeatedly during person's life), and about 2 million Americans and 300,000 Canadians suffer from epilepsy. There are many types and forms of seizures that range from a short episode of blank staring to convulsions -- and most seizures only last from 1-3 minutes or less.

### Things to watch for…

**Victim falls to floor and shakes or twitches in the arms, legs or body for a minute or longer**

**Blank staring or vacant expression and minor twitching of the face or jerking of the hand** (usually a mild epileptic seizure)

**Loss of body fluids or functions** (drooling, may pee or poop)

**No memory of what happened, confusion**

### What to do…

- Have someone call for an ambulance, especially if victim was poisoned or injured or if seizure lasts more than 3-5 minutes.
- Stay calm… you can't stop the convulsion or seizure!
- DO NOT put anything between victim's teeth or in their mouth!
- Move things that could hurt or fall on victim.
- Put something soft under victim's head, if possible.
- When convulsion or seizure is over, help roll victim on to their side to keep an open airway.
- Look for any other injuries and keep checking **ABCs… Airway, Breathing & Circulation**.
- Stay with victim until help arrives and try to calm them down.

### If victim is epileptic:

Ask if the victim takes any medications for seizures and help him/her take them according to the instructions.

*Also may want to review TERRORISM (in Section 2) for information, signs & symptoms, and treatment for several chemical agents that may cause convulsions.*

# What are <u>YOU</u> gonna do about...
## DIZZINESS & FAINTING?

## DIZZINESS

Dizziness is primarily a symptom and is usually combined with nausea (feel sick to the stomach), sweating, and a feeling of some kind of movement that really isn't there.

**Things to watch for...**
> **If dizzy feeling does not pass quickly or is really bad**
> **Fainting or passing out**
> **Vapors, mist or strange smells**

**What to do...**
- Have victim sit or lie down and close their eyes or focus on a nearby object that is not moving.
- Tell victim to try to keep their head still.

## FAINTING

Fainting is a temporary loss of consciousness (passing out) and may indicate a more serious condition. It's usually caused due to a lack of oxygenated blood to the brain. Be aware several types of injuries could cause fainting.

**Things to watch for...**
> **Visible injuries like bleeding from the ears or a bite or sting**
> **Pupils are enlarged or very small** (if different sizes, it could be a stroke)
> **Vapors, mist or strange smells**
> **Bluish color of skin, lips, fingertips or nails** (may not be getting air - *see BREATHING PROBLEMS*)

**What to do...**
- If victim is still passed out, put victim on their side to keep an open airway.
- Once victim is awake, gently roll them onto their back.
- Prop feet and the lower part of legs up with pillows or something (only if victim is not hurt).
- Loosen any tight clothing, especially around neck and waist.
- Check **ABCs... Airway, Breathing & Circulation**.
- Make sure victim rests before trying to get up.
- If necessary, contact doctor if symptoms persist.

# What are <u>YOU</u> gonna do about...
## DROWNING?

Things to watch for...
> **Signs of breathing**
> **Bluish color of skin, lips, fingertips or nails** (may not be getting air)
> **Pulse**

What to do...

- Have someone call for an ambulance.
- Once victim is out of the water, check **ABCs... Airway, Breathing & Circulation**.
- Check to see if there are any injuries or objects in mouth.
- If victim is not breathing or has no pulse, begin Rescue Breathing and/or CPR. *(see BREATHING PROBLEMS for Rescue Breathing and HEART PROBLEMS for CPR)*
- Once victim starts breathing on their own, cover with a blanket or dry towels to keep warm and have them lay on their side for a while.
- Stay with victim at all times until medical help arrives.

# What are <u>YOU</u> gonna do about…
## EAR INJURIES?

The ear is made up of 3 parts - sound waves are collected by the **outer** ear with its funny twists and turns. Sound then travels to the **middle** ear, which has three tiny bones behind the eardrum and is filled with air. From there, sounds (now vibrations) move on to the **inner** ear where the vibrations are turned into electrical signals and sent to the brain.

It's actually a much more involved process than listed above, but many types of ear injuries - from infections to a ruptured eardrum by a blast - will most likely require a medical professional.

Below are two minor ear-related injuries and tips on noise-induced hearing loss that could occur in some types of natural or man-made disasters.

## EAR INFECTIONS (OUTER / MINOR)

**Things to watch for…**
> **Itchy or tickly feeling in ear canal**
> **Pain when moving or tugging ear**
> **Inflamed or swollen ear canal**
> **Watery or smelly fluid may drain from ear**
> **Fever or dizziness**

**What to do…**
- Use ear drops or gently flush ear with 3% hydrogen peroxide.
- Keep ear canal dry when swimming or showering (use ear plugs or shower cap [in shower]).
- If ear feels plugged, try opening it by yawning, chewing gum or drinking warm teas or soup.
- Boost immune system to help fight infection (like taking astragalus, Vitamin C, garlic, mushrooms, zinc, and a good multiple vitamin + mineral supplement, etc. - but check with doctor if taking prescription medications).
- If pain continues or fever gets above 103F [39C] or fluid drains from ear - visit your doctor or an Ear, Nose and Throat specialist.
- Middle and inner ear infections can lead to long-term problems if left untreated, especially for small children.

# FOREIGN OBJECT IN EAR

If something crawls in or gets stuck in the ear...

- Keep victim calm and ask them sit down with head tilted sideways.
- Use a flashlight to try to see object in ear...

  **if a bug** - turn ear up toward sun or flashlight - most bugs are drawn to light so it might crawl out on it's own

  **if a loose item** - tilt head and try to shake it out

  **if still in but see it** - IF the victim is calm and you see the item, gently try to remove with tweezers, but DO NOT do this if the victim is squirming or item is deep in ear - you could damage the eardrum!

- Get medical help if you're not successful or can't visually locate object (doctors have special tools they use for ears).


# NOISE-INDUCED HEARING LOSS (NIHL)

Millions of people are exposed to hazardous sound levels daily. Loud impulse noise (like an explosion) or loud continuous noise at work or play (like working in a nightclub or snowmobiling) damage the delicate hair cells of the inner ear and the hearing nerve.

Sometimes damage can be temporary - like after a concert when your ears ring for a bit then go back to normal. But repeated loud noise over your life or a massive impulse noise could lead to permanent damage - damage that cannot be reversed. Just be aware certain types of disasters like tornadoes, a terrorist's bomb, or even hurricanes can cause a form of hearing loss.

**Things to watch for...**

**Bombs, tornadoes, power tools, loud music, motorcycles, etc.**
**Having to shout to be heard over noise (probably too loud)**
**Ringing or buzzing in ears**

**What to do...**

- Avoid loud situations or at least wear ear plugs or ear muffs.
- Have hearing tested yearly (esp. if you work around loud noises).
- Protect childrens' ears. (Note: using cotton is not enough.)

*To learn more about NIHL, visit the National Institute on Deafness and Other Communication Disorders' web site at www.nidcd.nih.gov (click on "Hearing, Ear Infections and Deafness") or call the NIHL Information Clearinghouse at 1-800-241-1044 or TTY: 1-800-241-1055*

# What are <u>YOU</u> gonna do about...
## Eye Injuries?

**Things to watch for...**
      **Severe or constant pain or burning**
      **Object stuck in the eye** (like a piece of metal or glass)
      **Redness and swelling**
      **Blurry vision, trouble keeping eye open or light sensitive**
      **Vapors or fumes in the air**
      **If injury is from a chemical, make a note of the name**
          **for Poison Control if possible**

**What to do...**
- Avoid rubbing the eye since this can cause more damage.
- Have victim sit down with their head tilted backwards.
- Wash hands before touching eye area!

**If the injury is a <u>loose</u> foreign object:**
- Gently separate the eyelids to see if you can locate a foreign object - can try removing it by wiping gently with damp tissue.
- Ask victim if he/she wears contact lenses, and if so, ask him or her to remove them.
- Have victim lean over sink or lie on back, hold eye open, and gently flush eye with lukewarm water or a saline solution.
- Get medical help if you are not successful!

**If there is an object sticking out of the eye:**
- Put thick soft pads around the object that is sticking out.
- DO NOT try to remove or press on the object!
- Carefully wrap with a roller bandage to hold thick pads around the object.
- Get medical attention immediately!

**If injury is from a blow to the eye:**
- Apply an icepack to reduce pain and swelling.
- Seek medical attention if damage to eye or blurred vision.

**If the injury is from a chemical:**
- Call your local Poison Control Center (or 1-800-222-1222 in the U.S.) and have name of chemical handy, if possible.

*Continued on next page...*

- If victim is wearing contact lenses, <u>ask Poison Control if they should be removed and whether to keep or dispose of them</u>! If okay to take out, ask victim to remove (if possible).
- Have victim lean over sink, lie down, or get in shower - hold eye(s) open, and gently flush with lukewarm water for at least 15 minutes. (If only one eye has chemical in it, make sure head is turned so it doesn't pour into the other eye!)
- Tell victim to roll eyeball(s) around while flushing to wash entire eye.
- DO NOT press or rub the eyes!
- May want to cover eyes with clean dressing and bandages but ask Poison Control or check label on bottle. For example, if chemical is mustard gas (sulfur mustard) you should <u>not</u> cover eyes ... but do wear shades to protect them.
- Get medical attention immediately!

**Things you should NOT do...**

- DO NOT try to remove an object that is stuck into the eye!
- DO NOT try to remove their contacts (if any)... let the victim do it!
- DO NOT try to move the eyeball if it comes out of the socket!

**Things you SHOULD do...**

- Protect your eyes with safety glasses or goggles when playing sports or working with tools or chemicals ... and wear shades during the day (to help reduce UV exposure).
- When an eye injury occurs, have an ophthalmologist (an eye physician <u>and</u> surgeon) examine it as soon as possible. You may not be realize how serious an injury is at first.

*Also may want to review TERRORISM (in Section 2) for information, signs & symptoms, and treatment on several biological and chemical agents that may cause eye injuries or discomfort.*

# What are <u>YOU</u> gonna do about...
## HEAD, NECK OR SPINE INJURIES?

**Things to watch for...**

> **Convulsions or seizures**
> **Intense pain in the head, neck or back**
> **Bleeding from the head, ears or nose**
> **Blurry vision**
> **Tingling or loss of feeling in the hands, fingers, feet or toes**
> **Weird bumps on the head or down the spine**
> **Shock** (pale, cold or clammy, drowsy, weak or rapid pulse, etc.)

**What to do...**

- Do not try to move victim unless they are in extreme danger and support the victim's head and neck during movement.

- Have someone call an ambulance immediately!

- Check to see if victim is alert and check **ABCs**... **Airway, Breathing & Circulation** ... and if you need to give them Rescue Breathing or CPR... DO <u>NOT</u> tilt their head back! *(see BREATHING PROBLEMS for Rescue Breathing and HEART PROBLEMS for CPR)*

- Try to control any bleeding using direct pressure. *(see BLEEDING)*

- If victim is passed out, hold their head gently between your hands while waiting for help to arrive. This will keep them from moving suddenly when/if they wake up.

# What are <u>YOU</u> gonna do about...
## HEART PROBLEMS?

**Heart attacks** can kill and most victims die within 2 hours of the first few symptoms. Most people deny they are having a heart attack - even if they have chest pains and shortness of breath... but DON'T take any chances! These are your body's warning signs, so pay attention! A heart attack can lead to <u>Cardiac Arrest</u>.

**Cardiac arrest** means that the heart stops beating and causes victim to pass out followed by no sign of breathing and no pulse.

**Cardiopulmonary resuscitation (CPR)** is used to help pump oxygenated blood through the body to the brain until the medical experts arrive. When you combine **CPR** and **Rescue Breathing**, you are giving the victim better odds of surviving since you help supply more oxygen to vital organs. However, if **CPR** is not done correctly, there's a chance of injuring a victim internally... especially on the elderly, children and infants.

Please realize the primary step during **CPR** is doing the compressions to keep blood flowing. If you have not been trained or don't feel comfortable doing **Rescue Breathing** - at least do the compressions. You don't even have to stop and check for a pulse - look and listen for signs and keep pumping!

The Red Cross teaches many First Aid courses, including CPR, so please contact your local Chapter and ask about their courses someday. *(See RED CROSS FIRST AID SERVICES & PROGRAMS at beginning of this section)*

## HEART ATTACK

**Things to watch for...**
> **Chest pain that can spread to the shoulder, arm, or jaw**
> **Shortness of breath or trouble breathing**
> **Strange pulse** (faster or slower than normal or sporadic)
> **Pale or bluish skin color**

**What to do...**
- Tell victim to STOP what they're doing and sit down and rest.
- Call for an ambulance immediately!
- Loosen any tight clothing, especially around neck and waist.
- Ask victim if they are taking any prescribed medicines for their heart... and if they do, have them take it!
- Take a couple of pure aspirin, if available.
- Watch victim's breathing and be prepared to give CPR. *(next pg)*

# CARDIAC ARREST (GIVING CPR)

Main thing is do compressions to keep blood moving until help arrives. Be aware steps below include special instructions for CHILDREN or INFANTS!

**Things to watch for...**
> **Not responding or passed out**
> **Not breathing and no pulse**
> **Broken bones or chest, head, neck or spine injuries**

**What to do...**
- Call for an ambulance immediately!
- Check **ABCs... Airway, Breathing, & Circulation**.
- Tilt head all the way back and lift chin. (Be careful with child's or infant's head... just tilt head a little bit!)
- Watch chest, listen, and feel for breathing for about 5 seconds.

**If victim is NOT breathing begin Rescue Breathing**...
- Pinch victim's nose shut.
- Open your mouth wide to make tight seal around victim's mouth *For INFANT - cover both mouth <u>and</u> nose with your mouth!*
- Give victim 2 slow breaths to make their chest rise.
- Watch chest, listen, and feel for breathing for a few seconds.

**To begin CPR**
- Find hand position in center of chest over breastbone – see illustrations on next 3 pages...
  FOR <u>ADULTS</u> – *[see illustration 3-1 on page 170]*
  FOR <u>CHILDREN</u> – *[see illustration 3-2 on page 171]*
  FOR <u>INFANTS</u> – *[see illustration 3-3 on page 172]*
- Begin chest compressions using the following guidelines...
  <u>ADULTS</u> – Using **both** hands, compress chest 15 times in 10 seconds.
  <u>CHILDREN</u> – Using **one** hand, compress chest 5 times in 3 seconds.
  <u>INFANTS</u> – Using **2 fingers**, compress chest 5 times in 3 seconds
- Breathe into the victim...
  <u>ADULTS</u> – Give 2 slow breaths
  <u>CHILDREN & INFANTS</u> – Give 1 slow breath
- Repeat chest compressions and breathing until victim recovers or ambulance arrives!
- If victim recovers (starts breathing and pulse resumes), then turn victim onto their side to keep the airway open

# CPR Position for Adults

*Illustration 3-1*

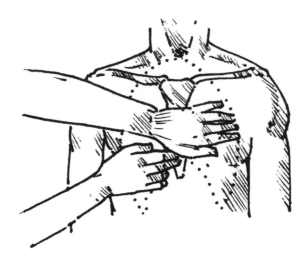

1. Find hand position

2. Position shoulders over hands. Compress chest 15 times.

# CPR POSITION FOR CHILDREN

*Illustration 3-2*

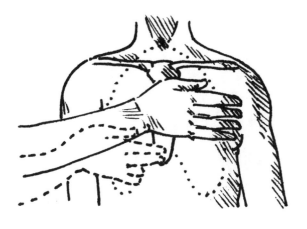

1. Find hand position

2. Position shoulder over hand. Compress chest 5 times.

# CPR Position for Infants

*Illustration 3-3*

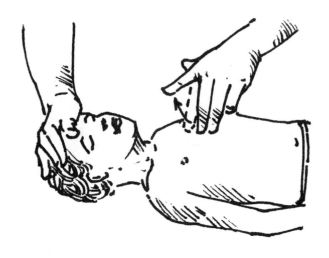

1. Find finger position

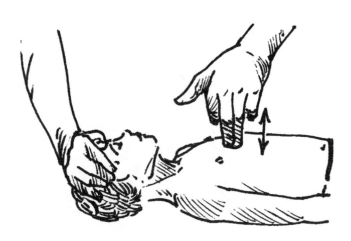

2. Position hand over fingers. Compress chest 5 times.

# What are <u>YOU</u> gonna do about...
## HEAT-RELATED ILLNESSES?

There are two major types of heat illness – **HEAT EXHAUSTION** and **HEAT STROKE.**  If heat exhaustion is left untreated it can lead to heat stroke. Both conditions are serious, however, **heat stroke** is a <u>major</u> medical emergency and getting the victim's body temperature cooled down is more critical than getting fluids in their body.  Heat stroke can lead to death.

**Things to watch for...**

<u>Heat Exhaustion</u>
> **Cool, clammy, or pale skin**
> **Light-headed or dizzy and weak**
> **Racing heart**
> **Sick to the stomach (nausea)**
> **Very thirsty or heavy sweating (sometimes)**

<u>Heat Stroke</u>
> **Very hot and dry skin**
> **Light-headed or dizzy**
> **Confusion, drowsiness or fainting**
> **Rapid breathing and rapid heartbeat**
> **Convulsions, passes out or slips into a coma**

**What to do...**

- Get victim to a cool or shady place (out of the sun) and rest.
- Loosen clothing around waist and neck to improve circulation, and remove sweaty clothes.
- Cool down victim's body - put wet cloths on victim's face, neck and skin and keep adding cool water to cloth... or if outdoors, use hose or stream. Also, fan the victim!
- Have victim drink <u>cool water</u>!  (NO alcohol – it dehydrates!)

**If victim refuses water, pukes or starts to pass out:**

- Call for an ambulance.
- Put victim on their side to keep airway open.
- Keep cooling down their body by placing ice or cold cloths on wrists, neck, armpits, and groin area (where leg meets the hip)!
- Check victim's **ABCs... Airway, Breathing, & Circulation**.
- Stay with victim until medical help arrives.

Remember, **HEAT STROKE** is a medical emergency and can cause victim to slip into a coma -- getting the victim's body temperature cooled down is more important than getting fluids in their body!

# What are <u>YOU</u> gonna do about…
## INFECTION?

It is important to be very careful and protect yourself against the spread of disease or infection when caring for an open wound or a wound that is bleeding. *(Review TIPS ON REDUCING THE SPREAD OF GERMS OR DISEASES at beginning of this section)*

## <u>INFECTION</u>

Germs are the main cause of an infection so whenever you perform first aid on anyone (including yourself), there is always a chance of spreading germs or diseases. All injuries - from tiny cuts to massive wounds - <u>must</u> be cleaned immediately to reduce the chances of infection! And keep cleaning a wound until it is completely healed.

Things to watch for…
> **Sore or wound is red and swollen or has red streaks**
> **Sore or wound is warm or painful**
> **Wound may open and ooze clear or yellowish fluid**
> **Fever or muscle aches or stiffness in the neck**

What to do…
- ALWAYS wash your hands before <u>and</u> after caring for a wound… even if it's your own!
- Immediately wash minor wounds with soap and water or rinse with hydrogen peroxide.
- Cover wound with sterile bandage or gauze -- best to clean and change bandage daily.
- Use an antibiotic cream or gel to help disinfect the wound and kill germs. *(Note: don't use cream or gel if wound is from a bite since it might seal wound and keep it from draining.)*
- Boost immune system to help fight infection (like taking astragalus, Vitamin C, garlic, mushrooms, zinc, and a good multiple vitamin + mineral supplement, etc. - but check with doctor if taking prescription medications).
- If infection gets worse, you may want to see a doctor.

# What are <u>YOU</u> gonna do about...
## POISONING?

Please make sure you have the local **Poison Control Center** phone number near a telephone since many poisonings can be cared for without the help of ambulance personnel. The people who staff Poison Control Centers (PCC) have access to information on most poisonous substances and can tell you what care to give to counteract the poison.

**POISON CONTROL CENTER # 1-800-222-1222 (U.S. only)**

**Internet:** www.1-800-222-1222.info

**If outside U.S., write in local Poison Control Centre phone # below:**

_____

## POISON - ABSORBED THROUGH THE SKIN

*Review TERRORISM (in Section 2) for information, signs & symptoms, and treatment on several poisonous biological and chemical agents that could be absorbed through skin.*

**Things to watch for...**
> **Reddened skin or burns**
> **Poison on skin, clothing or in the area**
> **Bites or marks from insect or animal** *(see BITES & STINGS)*
> **Possible Allergic Reaction -** Pain, discoloration or redness at site, trouble breathing, signs of shock (pale, cold, drowsy, etc.)

**What to do...**
- Be aware and make sure it is safe...then ask what happened.
- Move victim to safety (away from poison), if necessary.
- Find the container (if any) or name of the poison and call local Poison Control Center or an ambulance.
- Carefully remove clothing that may have poison on it and store in a bag someplace safe so people or animals won't touch them by accident! Ask authorities how to dispose of bag too.
- Flood skin with running water (hose or faucet) for 10 minutes.
- Wash area gently with soap and water. *(Note: some chemical agents don't suggest using soap so may want to ask.)*
- Monitor victim's breathing and watch for any allergic reactions.

# POISON - INHALED BY BREATHING

*Review TERRORISM (in Section 2) for information, signs & symptoms, and treatment on several poisonous biological and chemical agents that could be inhaled from humans' or critters' wet or dried body fluids, from soil, or from powders, gas, mist, or vapors.*

**Things to watch for...**
>**Strong odors or fumes**
>**Find the source of the odor or fumes** (be aware of threat)
>**Difficulty in breathing or dizzy**

**What to do...**
- Be aware and make sure it is safe...then ask what happened.
- Get victim out to fresh air.
- Avoid breathing fumes and open windows and doors (if safe).
- Call the Poison Control Center or an ambulance.
- If victim isn't breathing consider doing Rescue Breathing - but ONLY if sure poison cannot be spread person to person. *(see TERRORISM then BREATHING PROBLEMS)*


# POISON - POISONOUS PLANTS (IVY, OAK, & SUMAC)

The most common poisonous plants found in Canada and the lower 48 states in the U.S. include:

**Poison ivy** - can grow as a shrub or vine and is found across most of Canada and the U.S. It has white or cream-colored berries (or flowers in Spring) and the leaves usually come in leaflets of three to a stem but vary in color, size, shape and texture around the world. *[see illustration 3-4 on page 178]*

**Poison oak** - can grow as a shrub or vine and is found throughout the West and Southwest (very common in Oregon and California). It also varies widely in shapes and colors but usually has the distinctive shape of an oak leaf and red fuzzy berries. The leaves usually come in leaflets of three to a stem but can be in groups of five or more. It is best to learn what it looks like where you live. *[see illustration 3-5 on page 178]*

**Poison sumac** - is a tall shrub or small tree and mostly lives in standing water (like swamps and peat bogs). It has whitish green berries and bright green, pointy leaves that grow 6 to 12 leaves in pairs along both sides of each stem plus one leaf on each tip. *[see illustration 3-6 on page 179]*

All 3 of these plants have a sap called **urushiol** [oo-roo-she-ol] which is a sticky, colorless oil that stains things black when exposed to air. This sap is in the leaves, berries, stems and roots of all 3 types and can stick around a long time if the sap stays dry (which is why you need to rinse off whatever gets exposed - yourself, pets, clothing, shoes and laces, tools or camping stuff.) Never burn these plants since smoke can carry the oil and irritate skin, nose or lungs.

The rash is caused by your body's reaction to this oil and can show up as quickly as a few hours or take several days ... or never, in some cases, since some people don't have reactions to urushiol. It can irritate pets' skin too.

Note: the rash itself is <u>not</u> contagious but the OIL (urushiol) is what is transferred by hands, fingers, clothing or fur!! Make sure you wash your hands after touching the rash to avoid spreading any urushiol!

**Things to watch for...**
> **Inflamed red rash**
> **Extremely itchy skin or burning feeling**
> **Blisters**
> **Swelling or fever**
> **Allergic reactions** (weakness, dizziness, swelling in mouth or lips, trouble breathing or swallowing)

**What to do...**
- It is CRITICAL to wash affected area thoroughly with soap and running water then apply rubbing alcohol using cotton balls on area to remove any excess oil as quickly as possible.
- Make sure you immediately and carefully remove and wash any clothing and/or shoes that got exposed to the poison. *(And remember to wash pets, tools or anything else exposed to oil!)*
- If a rash or sores develop, use calamine lotion or baking soda paste several times a day on area.
- Take an antihistamine to reduce reaction (but read label first).
- If condition gets worse or spreads onto large areas of body, eyes or face, see a doctor.

**To relieve pain from poison ivy, oak, or sumac:**

**Baking soda** - Make a paste using 3 parts baking soda with 1 part water and apply on rash.

**Clay mudpack** - Leave clay on until it dries and shower it off. DO <u>NOT</u> use this method if <u>any</u> skin is broken or cracked to prevent infection!

**Jewelweed** - Studies indicate a plant that grows near poison ivy called jewelweed (has tiny, orange-yellow, horn-shaped flowers with reddish or white spots) can be used directly on area that brushed against ivy. Crush the juicy stems in your hands and apply to the area - even if there is no rash yet since it may help reduce inflammation. *[see illustration 3-7 on page 179]*

*Illustration 3-4*
**Poison Ivy**

*Illustration 3-5*
**Poison Oak**

*Illustration 3-6*
**Poison Sumac**

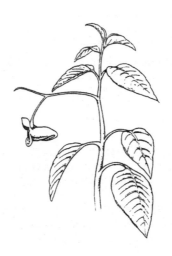

*Illustration 3-7*
**Jewelweed**

# POISON - SWALLOWED

*See TERRORISM (in Section 2) for information, signs & symptoms, and treatment on several poisonous biological and chemical agents that could be swallowed from eating or drinking something contaminated.*

**Things to watch for...**
> **Burns on the mouth, tongue and lips**
> **Stomach pains**
> **Open medicine cabinet; spilled or open containers**
> **Difficulty breathing**
> **Convulsions or seizures**
> **Weakness or dizziness**
> **Passed out**

**What to do...**

- Find out exactly what, how much, and how long ago it was swallowed.

- Call Poison Control Center or an ambulance and have bottle or container handy (if possible).

- NEVER give victim anything to eat or drink unless told to do so by Poison Control Center or a Medical professional!!

- If victim throws up, lay them on their side to keep airway open. Save a sample of the vomit IF the poison is unknown so the hospital can try to identify it.

- If victim isn't breathing consider doing Rescue Breathing - but ONLY if sure poison cannot be spread person to person. *(see TERRORISM then BREATHING PROBLEMS)*

It's a good idea to keep a few 1 ounce bottles of SYRUP OF IPECAC (pronounced ip'- î - kak) in your First Aid Kit and <u>use ONLY on the advice of a Medical professional or the Poison Control Center</u>!

Syrup of Ipecac is sold by most pharmacies without a prescription and used to induce vomiting (makes you puke) -- again, use only if instructed to do so.

# What are <u>YOU</u> gonna do about...
# SARS (Severe Acute Respiratory Syndrome)?

Severe acute respiratory syndrome (SARS) is a viral respiratory illness caused by a coronavirus. These types of viruses have occasionally been linked to pneumonia in humans (esp. people with weakened immune systems) and can cause severe disease in animals.

SARS appears to spread by close person-to-person contact primarily by touching people or things contaminated with bodily fluids (like droplets from coughing or sneezing) -- then touching your eyes, nose, or mouth. All body fluids (sweat, spittle, vomit, pee, etc.) may be infectious and may survive on objects for days. It's possible SARS can be spread more broadly through the air or other ways currently unknown. The best defense is good hygiene.

**Things to watch for...**
    **Fever -** higher than 100.4F (> 38.0C)
    **Possible early symptoms** - chills, headache, sore throat, general
        discomfort or body aches, mild respiratory symptoms, diarrhea
    **After 2 to 7 days, one or more respiratory symptoms appear** -
        dry cough, shortness of breath or trouble breathing, pneumonia

**If you or someone in your family might have SARS, you should...**
- Contact your health care provider as soon as possible!
- Cover mouth and nose with tissue when coughing or sneezing, wash hands often, and wear a surgical mask around others.

**If you or a family member has SARS and getting home care ...**
- Follow doctor's instructions and limit activities outside home.
- Everyone wash hands often or use alcohol-based hand rubs.
- SARS patient should cover mouth and nose with tissue when coughing or sneezing and wear a surgical mask around others.
- Clean counter or tabletops, doorknobs, bathroom fixtures, phones, etc. often with bleach and wear disposable gloves when cleaning.
- Don't share silverware, towels, or bedding & wash with hot water.
- Do above for 10 days <u>after</u> fever and symptoms have gone away.

*For more information about **SARS**, visit the Center for Disease Control's web site at <u>www.cdc.gov/ncidod/sars</u> or call CDC's Public Response Hotline at 1-888-246-2675 or 1-888-246-2857 (Español) or 1-866-874-2646 (TTY). Or visit Health Canada's web site at <u>www.sars.gc.ca</u> or call their public information line at 1-800-454-8302. Or visit the World Health Organization's site at <u>www.who.int/csr/sars/en/</u>*

---

# What are <u>YOU</u> gonna do about...
## SHOCK?

**Things to watch for...**
> **Pale, cold, and clammy skin**
> **Rapid heartbeat but weak pulse**
> **Quick and shallow breathing**
> **Dizziness or confusion**
> **Bluish color on lips and fingertips or nails**
> **Sick to their stomach or puking**
> **Intense thirst**

**What to do...**

- Call for an ambulance.
- Look for injuries and watch **ABCs... Airway, Breathing & Circulation**.
- Position victim using the following tips:
  <u>alert and awake</u> - place victim flat on their back with legs raised slightly
  <u>passed out or puking</u> - place victim on side to keep airway open
- Loosen any tight or restrictive clothing.
- Cover victim with a blanket or towel.
- Talk calmly to victim until help arrives (whether they are alert or not!)

# What are <u>YOU</u> gonna do about...
## A STROKE?

According to the American Stroke Association, over 600,000 Americans suffer strokes each year and about one-fourth of those victims die making stroke the 3rd leading cause of death in the U.S. Canada reports about 40,000-50,000 new strokes annually killing about 16,000 Canadians making it the 4th leading cause of death according to the Heart and Stroke Foundation of Canada.

A stroke (or "brain attack") occurs when oxygen and vital nutrients carried by blood are cut off causing brain cells to die. The reason it's cut off is because...

> ...a blood vessel is blocked in the neck or brain (by a blood clot or narrowing of an artery) -- called an **ischemic** stroke *(causes about 83% of strokes)*

... **or** ...

> ...a blood vessel bursts or leaks -- called **hemorrhagic** [hem-er-á-jik] stroke or bleeder *(causes 17% of strokes)*

**NOTE:** You only have 2 - 6 hours maximum to stop permanent brain damage from a stroke - so get to a hospital as quickly as possible (within 3 hours is best)! Most Emergency Rooms now have medicine to reduce the damaging long-term effects of a stroke - if medication is given in time!

**Things to watch for...**
> **Slurring or mumbling**
> **Loss of balance or stumbling**
> **Different sized pupils** (one pupil small and one pupil enlarged)
> **Loss of muscle control on one side of the body**
> **Severe headache**
> **Blurred or double-vision**
> **Shock** (pale, cold or clammy, drowsy, weak or rapid pulse, etc.)

**What to do...**
- Call for an ambulance.
- Get victim to lie back with head raised (put pillows or blanket blankets under head and shoulders so partially sitting up).
- Loosen any tight or restrictive clothing.
- See if there are any other injuries.
- If victim is drooling or having problems swallowing, place them on their side to keep the airway open.
- Stay with victim until medical help arrives.

# FACTS & INFORMATION ABOUT
# SUDDEN INFANT DEATH SYNDROME

Fedhealth and the Sudden Infant Death Syndrome (**SIDS**) Alliance want everyone to recognize that SIDS strikes without warning and affects families of all races and income levels.

Sudden Infant Death Syndrome is a medical disorder, which claims the lives of thousands of young children one day to one year of age. SIDS is the term used to describe infant deaths that remain unexplained after a thorough autopsy, death scene investigation and review of the medical history, ruling out other natural or unnatural causes of death.

## Did you know...

- … more children die of SIDS in a year than all children who die of cancer, heart disease, pneumonia, child abuse, AIDS, cystic fibrosis, and muscular dystrophy... combined?!

- … **African-American** babies are almost **twice** as likely to die of SIDS than white babies?! … and the rate of SIDS in the **Native-American** population is almost **3 times higher** than that of the population at large?!

- … more babies die of SIDS during the cold weather months?!

- … placing a baby on its back to sleep is best - doctors have found that a baby will NOT choke on spit-up or vomit so remember... "back is best"!

- … since the "Back to Sleep Campaign" was announced in 1992 the SIDS rate has decreased by 42%… the equivalent of sparing the lives of over 2,000 American babies a year?!

- … studies show *changing* a baby's sleeping position from his/her back can potentially increase the risk of SIDS **dramatically**?!

- … After 30 years of research, scientists still have not found a cause for SIDS. Although there are factors that may reduce the risks, there is no way to predict or prevent Sudden Infant Death Syndrome.

**SIDS is not contagious.**

**SIDS is not caused by immunizations.**

**SIDS is not caused by child abuse.**

**SIDS is no one's fault.**

# SUDDEN INFANT DEATH SYNDROME (SIDS) ALLIANCE
### *Advancing Infant Safety and Survival Across America*

The SIDS Alliance and Fedhealth are asking everyone to <u>PLEASE</u> educate child-care providers, baby-sitters, grandparents, and others about the SIDS risk reduction factors!

### REDUCE THE RISKS FOR SIDS: A CHECKLIST FOR NEW PARENTS

- ✔ Place your baby on the back to sleep at nighttime and naptime
- ✔ Use a firm mattress in a safety-approved crib
- ✔ Eliminate soft bedding from your baby's sleep area
- ✔ Keep your baby's face and head clear of blankets and other coverings during sleep
- ✔ Be careful not to overheat your baby
- ✔ Provide a smoke-free environment for your baby
- ✔ Educate baby sitters, grandparents and child care providers about SIDS risks
- ✔ And don't forget to enjoy your new baby!

### HELP FIGHT SIDS WITH THIS CHECKLIST AND A CHECK CALL 1-800-221-SIDS

The Sudden Infant Death Syndrome Alliance is a national, non-profit voluntary health organization uniting parents, caregivers, health professionals and researchers with government, business and community service groups concerned about the health of America's babies. The SIDS Alliance is a partner with the U.S. Public Health Service and the American Academy of Pediatrics in the Back to Sleep campaign, a nationwide infant health initiative aimed at reducing SIDS and infant mortality.

The SIDS Alliance funds medical research; offers emotional support nationally and through local Affiliate programs to families who have lost babies to SIDS; and supplies up-to-date information on SIDS to the general public, particularly new and expectant parents, through a nationwide, 24-hour toll free hotline **(1-800-221-7437)** and website **www.sidsalliance.org**.

### SIDS ALLIANCE ★ 1314 BEDFORD AVENUE, SUITE 210 ★ BALTIMORE, MD 21208

# Section 4

# Organizations' Emergency Contact Names & Numbers

# About the American Red Cross:

*Extracted from the American Red Cross Disaster Services web site:*

The mission of the American Red Cross Disaster Services is to ensure nationwide disaster planning, preparedness, community disaster education, mitigation, and response that will provide the American people with quality services delivered in a uniform, consistent, and responsive manner.

The American Red Cross responds to disasters such as hurricanes, floods, earthquakes, and fires, or other situations that cause human suffering or create human needs that those affected cannot alleviate without assistance. It is an independent, humanitarian, voluntary organization, not a government agency.

All Red Cross assistance is given free of charge, made possible by the generous contribution of people's time, money, and skills.

## Contacting Your Local American Red Cross:

There are a few different ways of finding your local Red Cross Chapter:

If you have access to the Internet, you can check their web site

### www.redcross.org

On the left hand side of the screen is a place to type in your zip code and click on FIND button

*... or ...*

You can browse through a list of local American Red Cross websites by clicking on link shown below zip code box on above web site

*... or ...*

Check your local telephone book in the white pages under BUSINESS LISTINGS for the American Red Cross!

Write in your local office here for future reference:

Local American Red Cross Address is:

_____

Telephone #: _____

# About the Federal Emergency Management Agency (FEMA):

*Extracted from FEMA's web site as of May 2004:*

On March 3, 2003 Federal Emergency Management Agency was transitioned into the U.S.'s Department of Homeland Security. *(also see APPENDIX A)*

FEMA's mission: to reduce loss of life and property and protect the nation's critical infrastructure from all types of hazards through a comprehensive, risk-based, emergency management program of mitigation, preparedness, response and recovery.

## FEMA Regional Offices

FEMA Region I
(serving CT, MA, ME, NH, RI, VT)
99 High Street
6th Floor
Boston, MA 02110
(617) 223-9540  FAX 617 223-9519
www.fema.gov/regions/i

FEMA Region II
(serving NJ, NY, PR, VI)
Suite 1307
26 Federal Plaza
New York, NY 10278-0001
(212) 680-3600  FAX 212 680-3681
www.fema.gov/regions/ii

FEMA Region III
(serving DC, DE, MD, PA,VA, WV)
One Independence Mall
615 Chestnut Street, 6th Floor
Philadelphia, PA 19106-4404
(215) 931-5608  FAX 215 931-5621
www.fema.gov/regions/iii

FEMA Region IV
(serving AL, FL, GA, KY, MS, NC, SC, TN)
3003 Chamblee Tucker Road
Atlanta, GA 30341
(770) 220-5200  FAX 770 220-5230
www.fema.gov/regions/iv

FEMA Region V
(serving IL, IN, MI, MN, OH, WI)
536 South Clark St., 6th Floor
Chicago, IL 60605
(312) 408-5500  FAX 312 408-5234
www.fema.gov/regions/v

FEMA Region VI
(serving AR, LA, NM, OK, TX)
Federal Regional Center
800 North Loop 288
Denton, TX 76209-3698
(940) 898-5399  FAX 940 898-5325
www.fema.gov/regions/vi

FEMA Region VII
(serving IA, KS, MO, NE)
2323 Grand Boulevard, Suite 900
Kansas City, MO 64108-2670
(816) 283-7061  FAX 816 283-7582
www.fema.gov/regions/vii

FEMA Region VIII
(serving CO, MT, ND, SD, UT, WY)
Denver Federal Center
Building 710, Box 25267
Denver, CO 80225-0267
(303) 235-4800  FAX 303 235-4976
www.fema.gov/regions/viii

FEMA Region IX
(serving AZ, CA, HI, NV, TERRITORIES [i.e. AMERICAN SAMOA, GUAM, etc.])
1111 Broadway, Suite 1200
Oakland, CA 94607
(510) 627-7100  FAX 510 627-7112
www.fema.gov/regions/ix

FEMA Region X
(serving AK, ID, OR, WA)
Federal Regional Center
130 228th St., SW
Bothell, WA 98021-9796
(425) 487-4600  FAX 425 487-4622
www.fema.gov/regions/x

# FEMA FOR KIDS:

FEMA has a fun web site for kids to learn about disasters, so if you have access to the Internet, check out their information and games for kids of all ages at www.fema.gov/kids *(and see pages 224-225 - more sites for kids!)*

# FEMA PARTNERS:

Emergency management is not the result of one government agency alone. FEMA works with many government, non-profit and private sector agencies to assist the public in preparing for, responding to, and recovering from a disaster. Together, these players make up the emergency response "team."

- Local Emergency Management Agencies
- State & Territory Emergency Management Offices
- National Emergency Management Organizations
- Federal-level Partners
- Partnerships with the Private Sector

In addition to these partners, FEMA's Global Emergency Management System (GEMS) provides access to a wide variety of emergency management and disaster related web sites. *Please visit FEMA's GEMS Links page at www.fema.gov/gems for a complete listing of partners.*

## DETAILED INFORMATION ABOUT FEMA PARTNERS:

### • LOCAL EMERGENCY MANAGEMENT AGENCIES

Even the largest, most widespread disasters require a local response. So local emergency management programs are the heart of the nation's emergency management system. FEMA supports them by offering training courses for emergency managers and firefighters, with funding for emergency planning and equipment, by conducting exercises for localities to practice their response, and by promoting ways to minimize disasters' effects. FEMA also builds partnerships with mayors, county boards, Tribal governments and other officials who share responsibility for emergency management.

Visit your City or County web site to see if they have a web link to your local Emergency Management, Emergency Services or Homeland Security Office or check the Blue Government pages in your city or county phone book.

### • STATE & TERRITORY EMERGENCY MANAGEMENT AGENCIES

Just like local EMAs above, every state emergency management agency is an integral part of the emergency management system. State and Territory offices coordinate federal, state, and local resources for mitigation, preparedness, response and recovery operations for citizens with support from FEMA and it's partners. *The next 4 pages list all State and U.S. Territory Emergency Management offices and agencies in alphabetical order.*

**Alabama** Emergency Management Agency
P. O. Drawer 2160
Clanton, AL 35046-2160
(205) 280-2200   FAX 205 280-2495
http://ema.alabama.gov

**Alaska** Division of Homeland Security and
Emergency Management
P. O. Box 5750
Fort Richardson, AK 99505-5750
(907) 428-7000  FAX 907 428-7009
www.ak-prepared.com

**American Samoa** Territorial Emergency
Management Coordination (TEMCO)
American Samoa Government
P. O. Box 1086
Pago Pago, American Samoa 96799
011 (684) 699-6415  FAX 011 684 699-6414

**Arizona** Div of Emergency Management
5636 East McDowell Road
Phoenix, AZ 85008
(602) 244-0504  FAX 602 231-6206
www.dem.state.az.us

**Arkansas** Dept of Emergency Management
P. O. Box 758
Conway, AR 72033
(501) 730-9750  FAX 501 730-9754
www.adem.state.ar.us

**California** Governor's Office of Emergency
Services
P. O. Box 419047
Rancho Cordova, CA 95741-9047
(916) 845-8500  FAX 916 845-8444
www.oes.ca.gov

**Colorado** Office of Emergency Management
Department of Local Affairs
15075 South Golden Road
Golden, CO 80401-3979
(303) 273-1622  FAX 303 273-1795
www.dola.state.co.us/oem/oemindex.htm

**Connecticut** Ofc of Emergency Management
Military Department
360 Broad Street
Hartford, CT 06105
(860) 566-3180  FAX 860 247-0664
www.ct.gov/oem

**Delaware** Emergency Management Agency
165 Brick Store Landing Road
Smyrna, DE 19977
(302) 659-3362  FAX 302 659-6855
www.state.de.us/dema

**District of Columbia** Emergency
Management Agency
2000 14th Street, NW, 8th Floor
Washington, DC 20009
(202) 727-6161  FAX 202 673-2290
http://dcema.dc.gov

**Florida** Division of Emergency Management
2555 Shumard Oak Blvd.
Tallahassee, FL 32399
(850) 413-9900  FAX 850 488-1016
www.floridadisaster.org

**Georgia** Emergency Management Agency
P. O. Box 18055
Atlanta, GA 30316-0055
(404) 635-7000  FAX 404 635-7205
www.state.ga.us/GEMA

**Guam** Homeland Security / Office of Civil
Defense
221B Chalan Palasyo
Agana Heights, Guam 96910
011 (671) 475-9600  FAX 011 671 477-3727
www.guamhs.org

**Hawaii** State Civil Defense
3949 Diamond Head Road
Honolulu, HI 96816-4495
(808) 733-4300  FAX 808 733-4287
www.scd.state.hi.us

**Idaho** Bureau of Homeland Security
4040 Guard Street, Building 600
Boise, ID 83705-5004
(208) 334-3460  FAX 208 334-2322
www.state.id.us/bds

**Illinois** Emergency Management Agency
110 East Adams Street
Springfield, IL 62701
(217) 782-7860  FAX 217 782-2589
www.state.il.us/iema

**Indiana** State Emergency Management
Agency
Room E-208
302 W. Washington Street
Indianapolis, IN 46204
(317) 232-3980  FAX 317 232-3895
www.in.gov/sema

**Iowa** Emergency Management Division
Department of Public Defense
Hoover State Office Building, Level A
Des Moines, IA 50319
(515) 281-3231  FAX 515 281-7539
www.state.ia.us/government/dpd/emd

**Kansas** Division of Emergency Management
2800 S.W. Topeka Boulevard
Topeka, KS 66611-1287
(785) 274-1409  FAX 785 274-1426
www.accesskansas.org/kdem

**Kentucky** Emergency Management
EOC Building
100 Minuteman Parkway
Frankfort, KY 40601-6168
(502) 607-1682  FAX 502 607-1614
http://kyem.dma.state.ky.us

**Louisiana** Homeland Security & Emergency
 Preparedness
7667 Independence Boulevard
Baton Rouge, LA 70804
(225) 925-7500  FAX 225 925-7501
www.loep.state.la.us

**Maine** Emergency Management Agency
72 State House Station
Augusta, ME 04333
(207) 626-4503  FAX 207 626-4499
www.state.me.us/mema

Commonweath of the Northern **Mariana
 Islands** Emergency Management Office
Office of the Governor
P. O. Box 10007
Saipan, Mariana Islands 96950
(670) 322-9529  FAX 670 322-9500
www.cnmiemo.org

National Disaster Management Office
Office of the Chief Secretary
P. O. Box 15
Majuro, Republic of **Marshall Islands** 96960
011 (692) 625-5181  FAX 011 692 625-6896

**Maryland** Emergency Management Agency
Camp Fretterd Military Reservation
5401 Rue Saint Lo Drive
Reistertown, MD 21136
(410) 517-3600  FAX 410 517-3610
Tollfree 1-877-MEMA-USA
www.mema.state.md.us

**Massachusetts** Emergency Management
 Agency
400 Worcester Road
Framingham, MA 01702-5399
(508) 820-2000  FAX 508 820-2030
www.state.ma.us/mema

**Michigan** Emergency Management Division
4000 Collins Road / P. O. Box 30636
Lansing, MI 48909-8136
(517) 333-5042  FAX 517 333-4987
www.michigan.gov/msp
*Click "Homeland Security" to get to EMD*

National Disaster Control Officer
Federated States of **Micronesia**
P. O. Box PS-53
Kolonia, Pohnpei - Micronesia 96941
011 (691) 320-8815  FAX 011 691 320-2785

**Minnesota** Homeland Security and
 Emergency Management (HSEM)
Department of Public Safety
444 Cedar Street, Suite 223
St. Paul, MN 55101-6223
(651) 296-2233  FAX 651 296-0459
www.hsem.state.mn.us

**Mississippi** Emergency Management Agency
P. O. Box 4501
Jackson, MS 39296-4501
(601) 352-9100  FAX 601 352-8314
Tollfree 1-800-442-MEMA (6362)
www.msema.org

**Missouri** Emergency Management Agency
P. O. Box 116
Jefferson City, MO 65102
(573) 526-9100  FAX 573 634-7966
www.sema.state.mo.us/semapage.htm

**Montana** Disaster and Emergency Services
P. O. Box 4789
Helena, MT 59604-4789
(406) 841-3911  FAX 406 841-3965
www.discoveringmontana.com/DMA/des

**Nebraska** Emergency Management Agency
1300 Military Road
Lincoln, NE 68508-1090
(402) 471-7421  FAX 402 471-7433
Tollfree 1-877-297-2368
www.nebema.org

**Nevada** Division of Emergency Management
2525 South Carson Street
Carson City, NV 89701
(775) 687-4240  FAX 775 687-6788
www.dem.state.nv.us

**New Hampshire** Department of Safety
Division of Fire Safety & Emergency Mgmt
Bureau of Emergency Management
33 Hazen Drive
Concord, NH 03305
(603) 271-2231  FAX 603 225-7341
www.nhoem.state.nh.us

**New Jersey** Office of Emergency
 Management
P. O. Box 7068
West Trenton, NJ 08628-0068
(609) 538-6050  FAX 609 538-0345
www.state.nj.us/njoem

**New Mexico** Office of Emergency
Management
P. O. Box 1628
Santa Fe, NM 87504-1628
(505) 476-9600  FAX 505 476-9650
www.dps.nm.org/emergency/index.htm

**New York** State Emergency Management
Office
Building 22, Suite 101
1220 Washington Avenue
Albany, NY 12226
(518) 457-2200  FAX 518 457-9995
www.nysemo.state.ny.us

**North Carolina** Emergency Management
116 West Jones Street
Raleigh, NC 27603
(919) 733-3867  FAX 919 733-7554
www.ncem.org

**North Dakota** Division of Emergency
Management
P. O. Box 5511
Bismarck, ND 58506-5511
(701) 328-8100  FAX 701 328-8181
www.state.nd.us/dem

**Ohio** Emergency Management Agency
2855 West Dublin-Granville Road
Columbus, OH 43235-2206
(614) 889-7150  FAX 614 889-7183
www.ema.ohio.gov

**Oklahoma** Department of Civil Emergency
Management
P. O. Box 53365
Oklahoma City, OK 73152-3365
(405) 521-2481  FAX 405 521-4053
www.odcem.state.ok.us

**Oregon** Emergency Management
P. O. Box 14370
Salem, OR 97309-5062
(503) 378-2911  FAX 503 373-7833
www.osp.state.or.us/oem/index.htm

**Palau** NEMO Coordinator
Office of the President
P. O. Box 100
Koror, Republic of Paulau 96940
011 (680) 488-2422  FAX 011 680 488-3312

**Pennsylvania** Emergency Management
Agency
2605 Interstate Drive
Harrisburg, PA 17110
(717) 651-2001  FAX 717 651-2040
www.pema.state.pa.us

**Puerto Rico** Emergency Management
Agency
P. O. Box 966597
San Juan, PR 00906-6597
(787) 724-0124  FAX 787 725-4244

**Rhode Island** Emergency Management
Agency
645 New London Avenue
Cranston, RI 02920-3003
(401) 946-9996  Fax 401 944-1891
www.riema.ri.gov

**South Carolina** Emergency Management
Division
1100 Fish Hatchery Road
West Columbia, SC 29172
(803) 737-8500  FAX 803 737-8570
www.scemd.org

**South Dakota** Division of Emergency
Management
118 West Capitol Avenue
Pierre, SD 57501
(605) 773-3231  FAX 605 773-3580
www.state.sd.us/dps/sddem/home.htm

**Tennessee** Emergency Management Agency
3041 Sidco Drive
Nashville, TN 37204
(615) 741-0001  FAX 615 242-9635
www.tnema.org

**Texas** Department of Public Safety
Division of Emergency Management
P. O. Box 4087
Austin, TX 78773-0001
(512) 424-2138  FAX 512 424-2444
www.txdps.state.tx.us/dem

**Utah** Division of Emergency Services
and Homeland Security
Room 1110, State Office Building
Salt Lake City, UT 84114
(801) 538-3400  FAX 801 538-3770
www.des.utah.gov

**Vermont** Emergency Management
Waterbury State Complex
103 South Main Street
Waterbury, VT 05671-2101
(802) 244-8721  FAX 802 244-8655
www.dps.state.vt.us/vem

**Virgin Islands** Territorial Emergency
Management Agency-VITEMA
2-C Contant, A-Q Building
St. Croix, VI 00820
(340) 774-2244  FAX 340 774-1491

**Virginia** Dept of Emergency Management
10501 Trade Court
Richmond, VA 23236-3713
(804) 897-6500  FAX 804 897-6626
www.vdem.state.va.us

State of **Washington** Military Department
Emergency Management Division
Building 20, M/S: TA-20
Camp Murray, WA 98430-5122
(253) 512-7000  FAX 253 512-7200
www.emd.wa.gov

**West Virginia** Office of Emergency Services
Building 1, Room EB-80
1900 Kanawha Blvd. East
Charleston, WV 25305-0360
(304) 558-5380  FAX 304 344-4538
www.state.wv.us/wvoes

**Wisconsin** Emergency Management
P. O. Box 7865
Madison, WI 53707-7865
(608) 242-3232  FAX 608 242-3247
http://emergencymanagement.wi.gov

**Wyoming** Office of Homeland Security /
  Emergency Management
122 West 25th Street
Herschler Bldg, 1st Floor East
Cheyenne, WY 82002
(307) 777-4663  FAX 307 635-6017
http://wyohomelandsecurity.state.wy.us

As of 12-May-2004 (per FEMA web site)
*Fedhealth verified links as of 22-May-2004*

## • NATIONAL EMERGENCY MANAGEMENT ORGANIZATIONS

**National Emergency Management Association** (NEMA) - membership includes State EM Directors.   **Internet:** www.nemaweb.org

**International Association of Emergency Managers** (IAEM) - membership includes local emergency managers.   **Internet:** www.iaem.com

## • FEMA'S FEDERAL-LEVEL PARTNERS

Numerous federal agencies and departments are partners in the nation's emergency management system. Before a disaster, they participate in training exercises and activities to help the nation become prepared. During a catastrophic disaster, FEMA coordinates the federal response, working with 27 federal partners and the American Red Cross to provide emergency food and water, medical supplies and services, search and rescue operations, transportation assistance, environmental assessment, and more.

The National Disaster Medical System is a partnership set up to provide emergency medical services in a disaster, involving FEMA, the Department of Health and Human Services, the Department of Defense, the Veterans Administration, as well as public and private hospitals across the country.

## • FEMA PARTNERSHIPS WITH THE PRIVATE SECTOR

FEMA encourages all sectors of society — from business and industry to volunteer organizations — to work together in disaster preparation, response and recovery. FEMA assists in coordinating activities of a variety of players, including private contractors, hospitals, volunteer organizations and area businesses. It's through these partnerships of people working together that communities are able to put the pieces back together. *(See APPENDIX B to learn about some volunteer opportunities.)*

# ABOUT THE CANADIAN RED CROSS:

*Extracted from the Canadian Red Cross Disaster Services web site:*

Canadian Red Cross helps people affected by emergencies and disasters -- situations ranging from a housefire to a flood that disrupts an entire region of the country. Following a disaster, Red Cross works with governments and other humanitarian organizations to provide for people's basic needs - food, clothing, shelter, first aid, emotional support and family reunification. The specific services offered will be based on the community's needs and the role that Red Cross has in the local disaster response plan.

All Red Cross assistance is provided free of charge and is made possible because of the generosity of financial donors and the volunteers who provide time and expertise.

## CONTACTING YOUR LOCAL CANADIAN RED CROSS:

There are a few different ways of finding your local Red Cross office:

If you have access to the Internet, you can check their national web site

### www.redcross.ca

Click on "Contact Us" then click on your "Zone" for list of offices

*... or call ...*

Canadian Red Cross
170 Metcalfe Street, Suite 300
Ottawa, Ontario K2P 2P2
Phone: 613.740.1900  Fax: 613.740.1911

To Donate by Phone: 1-800-418-1111

*... or ...*

Check your local telephone book in the white pages
for the Canadian Red Cross!

Write in your local office here for future reference:

Local Canadian Red Cross Address:

_____

Telephone #: _____

# ABOUT PUBLIC SAFETY AND EMERGENCY PREPAREDNESS CANADA

*Extracted from PSEPC and OCIPEP web sites as of May 2004:*

In December 2003, Prime Minister Paul Martin announced the Office of Critical Infrastructure Protection and Emergency Preparedness would be integrated into a new department, **Public Safety and Emergency Preparedness Canada**, to further secure the health and safety of Canadians.

The PSEPC portfolio includes emergency preparedness, crisis management, national security, policing and law enforcement, and border functions. *Due to space limitations this section just focuses on preparedness functions.*

Regional offices assist provincial and territorial Emergency Measures Organizations if an emergency escalates beyond their resource capabilities.

## PSEPC REGIONAL OFFICES

**PSEPC Headquarters**
Public Safety and Emergency Preparedness
  Canada
Communications Branch
340 Laurier Avenue West
Ottawa, ON  K1A 0P8
Tel: (613) 944-4875  Fax: (613) 998-9589
Urgent Matters: (613) 991-7000
communications@psepc-sppcc.gc.ca
PSEPC web site: www.psepc-sppcc.gc.ca
OCIPEP web site: www.ocipep.gc.ca

**Newfoundland and Labrador**
P. O. Box 668, Station 'C'
St. John's, NL  A1C 5L4
Tel: (709) 772-5522  Fax: (709) 772-4532

**Nova Scotia**
Suite 219, 21 Mount Hope Ave
Dartmouth, NS  B2Y 4R4
Tel: (902) 426-2082  Fax: (902) 426-2087

**Prince Edward Island**
134 Kent Street, 6th Floor
Charlottetown, PE  C1A 8R8
Tel: (902) 566-7047  Fax: (902) 566-7045

**New Brunswick**
P. O. Box 534
Fredericton, NB  E3B 5A6
Tel: (506) 452-3020  Fax: (506) 452-3906

**Québec**
Champlain Harbour Station, Room 350-1
901 Cap Diamant
Quebec, QC  G1K 4K1
Tel: (418) 648-3111  Fax: (418) 648-3165

**Ontario**
4900 Yonge Street, Suite 240
Toronto, ON  M2N 6A4
Tel: (416) 973-6343  Fax: (416) 973-2362

**Manitoba**
Suite 403, MacDonald Building
344 Edmonton Street
Winnipeg, MB  R3B 2L4
Tel: (204) 983-6790  Fax: (204) 983-3886

**Saskatchewan**
320 - 1975 Scarth Street
Regina, SK  S4P 2H1
Tel: (306) 780-5005  Fax: (306) 780-6461

**Alberta, Nunavut & Northwest Territories**
Suite 150, 10130-103 Street N.W.
Edmonton, AB  T5J 3N9
Tel: (780) 495-3005  Fax: (780) 495-3585

**British Columbia and Yukon**
P. O. Box 10,000
Victoria, BC  V8W 3A5
Tel: (250) 363-3621  Fax: (250) 363-3995

# PSEPC Mandates:

- to provide national leadership of a new, modern and comprehensive approach to protecting Canada's critical infrastructure -- the key physical and cyber components of the energy and utilities, communications, services, transportation, safety and government sectors
- to be the government's primary agency for ensuring national civil emergency preparedness -- for all types of emergencies.

## Emergency Preparedness in Canada:

Emergency preparedness in Canada is a shared responsibility.

- It is up to the individual to know what to do in an emergency.
- Local officials and response organizations handle local emergencies.
- Provincial and Territorial Emergency Measures Organizations manage large scale emergencies (prevention, preparedness, response and recovery) and support municipal and community response teams.
- Requests from provinces to the Government of Canada are managed through PSEPC's Regional and Headquarters offices. *(Please note there are many departments and agencies that work on prevention, response, recovery, and security in collaboration with PSEPC.)*

The following explains Canada's **Emergency Management System**:

### • Emergency Operations

Through the Government Emergency Operations Coordination Centre (GEOCC), PSEPC maintains an around the clock monitoring and information centre of actual, potential and imminent disasters. During major events, the GEOCC, with the help of emergency personnel from other departments, serves as the focal point for emergency government operations.

### • Financial Assistance Programs

PSEPC administers the Joint Emergency Preparedness Program (JEPP) and the Disaster Financial Assistance Arrangements (DFAA). JEPP funding ensures communities have response skills and equipment in place to deal with emergency situations of any type. DFAA shares the costs of responding to and recovering from disasters when the costs of doing so exceed the fiscal capacity of provincial and territorial governments.

### • Training And Education

The Canadian Emergency Preparedness College in Ontario provides training and education for first responders and emergency managers.

### • Public Information

Public awareness campaigns provide individuals with information needed to become better prepared for an emergency. PSEPC manages Emergency

Preparedness Week, which is held every May and is jointly delivered with provincial and territorial governments, municipalities, NGOs, volunteers, teachers and others. And PSEPC offers a number of communications products online, through Safe Guard partners, and in various publications.

# PSEPC PARTNERSHIPS:

## • DOMESTIC PARTNERSHIPS

Domestic partnerships work toward the common goals of protecting Canada's critical infrastructure and mitigating associated risks with all levels of government (federal, provincial, territorial and municipal), as well as the private and voluntary sectors. The PSEPC and it's six member agencies (e.g. RCMP, Canadian Security Intelligence Service, Canada Border Services Agency, etc.) are maximizing interagency cooperation.

During a major disaster or emergency, PSEPC works closely with various Government of Canada departments, agencies and Provincial and Territorial Emergency Measures Organizations (listed below in alphabetical order).

### PROVINCIAL & TERRITORIAL
### EMERGENCY MEASURES ORGANIZATIONS (EMOS)

**Alberta**
Emergency Management Alberta
Alberta Municipal Affairs
16th Floor, Commerce Place
10155 - 102nd Street
Edmonton, AB  T5J 4L4
Tel: (780) 422-9000  Fax: (780) 422-1549
Tollfree 310-0000  (in Alberta)
www.municipalaffairs.gov.ab.ca/ema

**British Columbia**
Provincial Emergency Program Headquarters
P. O. Box 9201 Station Prov. Govt.
Victoria, BC  V8W 9J1
Tel: (250) 952-4913  Fax: (250) 952-4888
www.pep.bc.ca

**PEP Central Region**
1255 - D Dalhousie Drive
Kamloops, BC V2C 5Z5
(250) 371-5240  Fax: (250) 371-5246

**PEP North East Region**
1541 South Ogilvie Street
Prince George, BC V2N 1W7
(250) 612-4172  Fax: (250) 612-4171

**PEP North West Region**
Suite 1B - 3215 Eby Street
Terrace, BC V8G 2X8
(250) 615-4800  Fax: (250) 615-4817

**PEP South East Region**
403 Vernon Street
Nelson, BC V1L 4E6
(250) 354-5904  Fax: (250) 354-6561

**PEP South West Region**
9800 - 140th Street
Surrey, BC V3T 4M5
(604) 586-2665  Fax: (604) 586-4334

**PEP Vancouver Island Region**
455 Boleskine Road
Victoria, BC V8Z 1E7
(250) 952-5848  Fax: (250) 952-4983

**Manitoba**
Manitoba Emergency Measures Organization
405 Broadway Avenue, Room 1525
Winnipeg, MB  R3C 3L6
Tel: (204) 945-4772  Fax: (204) 945-4620
Tollfree 1-888-267-8298
www.ManitobaEMO.ca

**New Brunswick**
New Brunswick Emergency Measures
  Organization
Department of Public Safety
P. O. Box 6000
Fredericton, NB   E3B 5H1
Tel: (506) 453-2133  Fax: (506) 453-5513
www.gnb.ca/cnb/emo-omu/index-e.asp

**Newfoundland and Labrador**
Emergency Measures Organization
Dept of Municipal and Provincial Affairs
P. O. Box 8700
St. John's, NL   A1B 4J6
Tel: (709) 729-3703  Fax: (709) 729-3857
www.gov.nf.ca/mpa/emo.html

**Northwest Territories**
Emergency Measures Organization
Dept. of Municipal and Community Affairs
Government of Northwest Territories
600, 5201 - 50th Avenue - Northwest Tower
Yellowknife, NT  X1A 3S9
Tel: (867) 873-7083  or  873-7785
Fax: (867) 873-8193
www.maca.gov.nt.ca/safety

**Nova Scotia**
Nova Scotia Emergency Measures
  Organization
P. O. Box 2581
Halifax, NS   B3J 3N5
Tel: (902) 424-5620  Fax: (902) 424-5376
www.gov.ns.ca/emo

**Nunavut**
Nunavut Emergency Management
Department of Community Government
  and Transportation
P. O. Box 1000, Station 700
Iqaluit, NU  X0A 0H0
Tel: (867) 975-5300  Fax: (867) 979-4221

**Ontario**
Emergency Management Ontario
Ministry of Public Safety and Security
77 Wellesley St. West, Box 222
Toronto, ON   M7A 1N3
Tel: (416) 314-3723  Fax: (416) 314-3758
www.mpss.jus.gov.on.ca/english/
  pub_security/emo/emo.html

**Prince Edward Island**
Emergency Measures Organization
134 Kent Street, Suite 600
Charlottetown, PE   C1A 8R8
Tel: (902) 368-6361  Fax: (902) 368-6362
www.gov.pe.ca/commcul/emo

**Québec**
Direction générale de la Sécurité civile et de
  la sécurité incendie
Ministère de la Sécurité publique
2525, boul. Laurier, 5e étage
Sainte-Foy, QC  G1V 2L2
Tel: (418) 644-6826  Fax: (418) 643-3194
www.msp.gouv.qc.ca

**Saskatchewan**
Saskatchewan Emergency Measures
  Organization  (SaskEMO)
100 - 1855 Victoria Avenue
Regina, SK   S4P 3V7
Tel: (306) 787-9563  Fax: (306) 787-1694
www.cps.gov.sk.ca/safety/emergency

**Yukon Territory**
Yukon Emergency Measures Branch
Department of Community Services
P. O. Box 2703, EMO
Whitehorse, YK  Y1A 2C6
Tel: (867) 667-5220  Fax: (867) 393-6266
www.gov.yk.ca/depts/community/emo

Per OCIPEP's site www.ocipep.gc.ca  and
www.emergencypreparednessweek.ca
*Fedhealth verified info & links:24-May-2004*

## • INTERNATIONAL PARTNERSHIPS

PSEPC and Government of Canada partners continue to build strong relationships with the United States in infrastructure protection departments and agencies, law enforcement and intelligence communities, and emergency management agencies.  Canada's also working with various countries on international issues of critical infrastructure vulnerabilities.

# INTERNATIONAL FEDERATION OF RED CROSS AND RED CRESCENT SOCIETIES

*The following was extracted from the IFRC web site:*

The **International Federation of Red Cross and Red Crescent Societies** is the world's largest humanitarian organization providing assistance without discrimination as to nationality, race, religious beliefs, class or political opinions.

Founded in 1919, the International Federation comprises 178 member Red Cross and Red Crescent societies, a Secretariat in Geneva and more than 60 delegations strategically located to support activities around the world. There are more societies in formation. The Red Crescent is used in place of the Red Cross in many Islamic countries.

The Federation's mission is **to improve the lives of vulnerable people by mobilizing the power of humanity**. Vulnerable people are those who are at greatest risk from situations that threaten their survival, or their capacity to live with an acceptable level of social and economic security and human dignity. Often, these are victims of natural disasters, poverty brought about by socio-economic crises, refugees, and victims of health emergencies.

The Federation's work focuses on four core areas: promoting humanitarian values, disaster response, disaster preparedness, and health and community care.

For more information, please visit IFRC on the Internet or contact:

**International Federation of Red Cross and Red Crescent Societies**

P. O. Box 372
CH-1211 Geneva 19
Switzerland
Telephone: (+41 22) 730 42 22
Fax: (+41 22) 733 03 95
**Internet: www.ifrc.org**
E-Mail: secretariat@ifrc.org

Or find your local National Society via the **Directory** link on the Internet. *(The online Directory has an alphabetic listing by country of all the Red Cross and Red Crescent Societies worldwide.)*

# APPENDIX A

# U. S. Department of Homeland Security

*Extracted from DHS' web site www.dhs.gov as of September 2003:*

## What is the Department of Homeland Security?

In the aftermath of the terrorist attacks against America on September 11th, 2001, President George W. Bush decided 22 previously disparate domestic agencies needed to be coordinated into one department to protect the nation against threats to the homeland.

The first priority of the **Department of Homeland Security** (DHS) is to protect the nation against further terrorist attacks. Component agencies will analyze threats and intelligence, guard borders and airports, protect critical infrastructure, and coordinate response on future emergencies.

DHS is also dedicated to protecting the rights of American citizens and enhancing public services, such as natural disaster assistance and citizenship services, by dedicating offices to these important missions.

## How is DHS organized?

DHS has Five Major Divisions, or "Directorates":

**Border and Transportation Security** (BTS) - maintains security of the U.S.'s borders and transportation systems (including waterways, ports, and terminals) to prevent the entry of terrorists and instruments of terror.

**Emergency Preparedness and Response** (EPR) - ensures the nation is prepared for, and able to recover from, terrorist attacks and natural disasters. *(Note: FEMA was transitioned under this Division as of March 3, 2003.)*

**Science and Technology** (S & T) - coordinates efforts in research and development, including preparing for and responding to the full range of terrorist threats involving weapons of mass destruction.

**Information Analysis and Infrastructure Protection** (IAIP) - merges the capability to identify and assess current and future threats to the nation, maps those threats against vulnerabilities, and provides timely warnings regarding physical and cyber threats through the Homeland Security Advisory System.

**Management** - handles DHS budget, management and personnel issues.

Some other key agencies folding into DHS are the U.S. Coast Guard, U.S. Secret Service, the Bureau of Citizenship and Immigration Services and the Office of State and Local Government Coordination.

# What is the Homeland Security Advisory System?

In March 2002, the **Homeland Security Advisory System (HSAS)** was implemented using color-coded "Threat Conditions" that increase or decrease based on reports from the Intelligence Community.

### <u>HSAS's "Threat Conditions" or "Threat Levels":</u>
SEVERE = **RED**  (Severe risk of terrorist attacks)

HIGH = **ORANGE**  (High risk of terrorist attacks)

ELEVATED = **YELLOW**  (Significant risk of terrorist attacks)

GUARDED = **BLUE**  (General risk of terrorist attacks)

LOW = **GREEN**  (Low risk of terrorist attacks)

Alerts and threat conditions can be declared for the entire nation, or for a specific geographic area or industry.

The public should stay current with news and alerts issued by officials ... and be aware, be prepared, and have a plan at <u>all</u> threat levels.

The **District of Columbia Emergency Management Agency (DCEMA)** developed and contributed the following "Terrorist Threat Advisory System" that mirrors the national Homeland Security Advisory System. The DCEMA's suggested precautions provide general guidance only to help organizations and families take actions best tailored for their needs.

*Please note, there are some protective measures for federal departments and agencies per DHS included here too.*

LOW *(Green)* - a **low risk** of terrorism. Routine security is implemented to preclude routine criminal threats.

<u>Residents are advised to</u>:
- Continue to enjoy individual freedom. Participate freely in travel, work, and recreational activities.
- Be prepared for disasters and family emergencies.
- Develop a family emergency plan.
- Keep recommended immunizations up-to-date.
- Know how to turn off power, gas, and water service to your house.
- Know what hazardous materials are stored in your home and how to properly dispose of unneeded chemicals.
- Support the efforts of your local emergency responders (fire fighters, law enforcement and emergency medical service).
- Know what natural hazards are prevalent in your area and what measures you can take to protect your family. Be familiar with local natural and technological (man made) hazards in your community.
- Volunteer to assist and support community emergency response agencies.

- Become active in your local Neighborhood Crime Watch program.
- Take a first aid or Community Emergency Response Team (CERT) class.

### *(Green)*  Business owners/managers are advised to:
- Develop emergency operations and business contingency plans.
- Encourage and assist employees to be prepared for personal, natural, technological, and homeland security emergencies.
- Conduct emergency preparedness training for employees and their families.
- Develop a communications plan for emergency response and key personnel.
- Conduct training for employees on physical security precautions.
- Budget for physical security measures.

### *(Green)*  Federal departments and agencies should consider:
- Refine and exercise planned Protective Measures.
- Ensure emergency personnel receive proper training on HSAS measures.
- Assess facilities for vulnerabilities and take measures to mitigate them.

GUARDED *(Blue)* - a **general risk** of terrorism with no credible threats to specific targets.

### In addition to all previously mentioned precautions, residents are advised to:
- Continue normal activities but be watchful for suspicious activities. Report suspicious activity to local law enforcement.
- Review family emergency plans.
- Avoid leaving unattended packages or briefcases in public areas.
- Increase family emergency preparedness by purchasing supplies, food, and storing water.
- Increase individual or family emergency preparedness through training, maintaining good physical fitness and health, and storing supplies.
- Monitor local and national news for terrorist alerts.

### In addition to all previously mentioned precautions, business owners and managers are advised to:
- Ensure that key leaders are familiar with the emergency operations and business contingency plans.
- Review, update, and routinely exercise functional areas of plans.
- Review and update the call down list for emergency response teams.
- Develop or review Mutual Aid agreements with other facilities and/or with local government for use during emergencies.
- Review physical security precautions to prevent theft, unauthorized entry, or destruction of property.
- Have you provided for:
  - Employee picture ID badges?

*... continued on next page ...*

- Background checks on all employees (as applicable)?
- Access control and locking of high security areas at all times?
- All security keys marked with "Do not Duplicate?"
- Surveillance Cameras?
- Backup power?
- An alarm system?

### *(Blue)*  In addition to all previously mentioned precautions, federal departments and agencies should consider:

- Check communications with designated emergency response or command locations.
- Review and update emergency response procedures.
- Provide public with information that would strengthen its ability to act appropriately.

ELEVATED *(Yellow)* - an **elevated risk** of terrorist attack but a specific region of the USA or target has <u>not</u> been identified.

### In addition to all previously mentioned precautions, residents are advised to:

- Continue normal activities, but report suspicious activities to the local law enforcement agencies.
- Network with your family, neighbors, and community for mutual support during a disaster or terrorist attack.
- Learn what critical facilities are located in your community and report suspicious activities at or near these sites.
- Contact local officials to learn about specific hazards in your community.
- Develop your family preparedness kit and plan and check the contents of your **Disaster Supplies Kit** *(see Section 1)*. Individual or family emergency preparedness should be maintained through training, good physical fitness and health, and storing food, water, and emergency supplies.
- Monitor media reports concerning situation.

### In addition to all previously mentioned precautions, business owners and managers are advised to:

- Announce Threat Condition **ELEVATED** to employees.
- Review vulnerability and threat assessments and revise as needed.
- Identify and monitor government information sharing sources for warnings and alerts.
- Update and test call down list for emergency response teams and key employees.
- Review, coordinate, and update mutual aid agreements with other critical facilities and government agencies.
- Establish and monitor more active security measures.
- Review employee training on security precautions (bomb threat procedures, reporting suspicious packages, activities, and people). Conduct communications checks to ensure contacts can be maintained.

*(Yellow)*  **In addition to all previously mentioned precautions, federal departments and agencies should consider:**

- Increase surveillance of critical locations.
- Coordinate emergency plans with nearby jurisdictions, as needed.
- Assess whether the precise characteristics of the threat require the further refinement of preplanned protective measures.
- Implement, as appropriate, contingency and emergency response plans.


**HIGH** *(Orange)* - credible intelligence indicates that there is a **high risk** of a local terrorist attack but a specific target has not been identified.

**In addition to all previously mentioned precautions, residents are advised to**:

- Resume normal activities but expect some delays, baggage searches, and restrictions due to heightened security at public buildings and facilities.
- Continue to monitor world and local events as well as local government threat advisories.
- Report suspicious activities at or near critical facilities to local law enforcement agencies by calling 9-1-1.
- Inventory and organize emergency supply kits and test emergency plans with family members. Reevaluate meeting location based on threat.
- Consider taking reasonable personal security precautions. Be alert to your surroundings, avoid placing yourself in a vulnerable situation, and monitor the activities of your children.
- Maintain close contact with family and neighbors to ensure their safety and emotional welfare.

**In addition to all previously mentioned precautions, business owners and managers are advised to**:

- Announce Threat Condition **HIGH** to all employees and explain expected actions.
- Place emergency response teams on notice.
- Activate the business emergency operations center if required. Establish ongoing liaison with local law enforcement and emergency management officials.
- Monitor world and local events. Pass on credible threat intelligence to key personnel.
- Ensure appropriate security measures are in place and functioning properly.
- Instruct employees to report suspicious activities, packages, and people.
- Search all personal bags, parcels, and require personnel to pass through magnetometer, if available.
- Inspect intrusion detection systems and lighting, security fencing, and locking systems.
- Inspect all deliveries and consider accepting shipments only at off-site locations.
- Remind employees to expect delays and baggage searches.
- Implement varying security measures *(listed on next page)*

*These measures incorporate a comprehensive list of security actions, some of which may need to be implemented at lower levels. They are designed to respond to the elevation to **HIGH** Risk (**Orange**) of terrorist attacks.*

## Varying Security Measures for Businesses - Little or No Cost Actions

- Increase the visible security personnel presence wherever possible.
- Rearrange exterior vehicle barriers (traffic cones) to alter traffic patterns near facilities.
- Institute/increase vehicle, foot, and roving security patrols.
- Implement random security guard shift changes.
- Arrange for law enforcement vehicles to be parked randomly near entrances and exits.
- Approach all illegally parked vehicles in and around facilities, question drivers and direct them to move immediately. If owner cannot be identified, have vehicle towed by law enforcement.
- Limit number of access points and strictly enforce access control procedures.
- Alter primary entrances and exits if possible.
- Implement stringent identification procedures to include conducting 100% "hands on" checks of security badges for all personnel, if badges are used.
- Remind personnel to properly display badges, if applicable, and enforce visibility.
- Require two forms of photo identification for all visitors.
- Escort all visitors entering and departing.
- X-ray packages and inspect handbags and briefcases at entry if possible.
- Validate vendor lists for all routine deliveries and repair services.

## Varying Security Measures - Actions That May Bear Some Cost

- Increase perimeter lighting.
- Remove vegetation in and around perimeters, maintain regularly.
- Institute a vehicle inspection program to include checking under the undercarriage of vehicles, under the hood, and in the trunk. Provide vehicle inspection training to security personnel.
- Conduct vulnerability studies focusing on physical security, structural engineering, infrastructure engineering, power, water, and air infiltration, if feasible.
- Initiate a system to enhance mail and package screening procedures (both announced and unannounced).
- Install special locking devices on manhole covers in and around facilities.

## (Orange) In addition to all previously mentioned precautions, federal departments and agencies should consider:

- Coordinate security efforts with federal, state and local law enforcement agencies, National Guard or other security and armed forces.
- Take additional precautions at public events, possibly considering alternative venues or cancellation.
- Prepare to work at an alternate site or with a dispersed workforce.
- Restrict access to a threatened facility to essential personnel only.

**SEVERE *(Red)*** - terrorist attack has occurred or credible and corroborated intelligence indicates that one is imminent (a **severe risk**). Normally, this threat condition is declared for a specific location or critical facility.

### In addition to all previously mentioned precautions, residents are advised to:

- Report suspicious activities and call 9-1-1 for immediate response.
- Expect delays, searches of purses and bags, and restricted access to public buildings.
- Expect traffic delays and restrictions.
- Residents should have **Disaster Supplies Kits** stocked and in place ready to go (medicines and medical supplies, glasses, contacts, important legal and financial papers) and emergency supplies kits (first aid kits, duct tape, blankets, non-perishable food, water) for sheltering in place, if requested to do so. *(see Section 1)*
- Take personal security precautions to avoid becoming a victim of crime or terrorist attack.
- Avoid participating in crowded optional public gatherings, such as sporting events and concerts. However, do not avoid going to public emergency gathering locations such as hospitals and shelters, if directed or necessary. These locations will have developed and initiated a strong security plan to protect the residents.
- Do not travel into areas affected by the attack or that are likely to become an expected terrorist target.
- Keep emergency supplies accessible and automobile fuel tank full.
- Be prepared to either evacuate your home or shelter-in-place on order of local authorities. *(see EVACUATION in Section 2)*
- Be suspicious of persons taking photographs of critical facilities, asking detailed questions about physical security or dressed inappropriately for weather conditions. Report these incidents immediately to law enforcement.
- Closely monitor news reports and Emergency Alert System (EAS) radio/TV stations.
- Assist neighbors who may need help.
- Ensure pets can be readied quickly for boarding or evacuation, if necessary.
- Avoid passing unsubstantiated information and rumors.
- Prepare to activate your personal Family Emergency Plan. *(see Section 1)*

### In addition to all previously mentioned precautions, business owners and managers are advised to:

- Announce Threat Condition **SEVERE** and explain expected actions.
- Deploy security personnel based on threat assessments.
- Close or restrict entry to the facility to emergency personnel only and restrict parking areas close to critical buildings.
- Maintain a skeleton crew of emergency employees.
- Deploy emergency response and security teams.
- Activate Operations Centers (if applicable).

*... continued on next page ...*

- Maintain close contact with local law enforcement, emergency management officials and business consortium groups (Chamber of Commerce, Board of Trade, etc…)
- Be prepared to implement mutual aid agreements with government and with other similar/neighboring businesses/industries.
- Provide security in parking lots and company areas.
- Report suspicious activity immediately to local law enforcement.
- Restrict or suspend all deliveries and mail to the facility. Emergency supplies or essential shipments should be sent to off-site location for inspection.
- Activate your business emergency/contingency plan.

*(Red)* **In addition to all previously mentioned precautions, federal departments and agencies should consider:**
- Increase or redirect personnel to address critical emergency needs.
- Assign emergency response personnel and pre-position and mobilize specially trained teams or resources.
- Monitor, redirect, or constrain transportation systems.
- Close public and government facilities not critical for continuity of essential operations, especially public safety.

For more information about the **District of Columbia Emergency Management Agency**'s Homeland Security Terrorist Threat Advisory System, please visit http://dcema.dc.gov (link under "Information" header)

Also, the American Red Cross has developed a complementary set of guidelines for Individuals, Families, Neighborhoods, Schools and Businesses (in English and Spanish) explaining the Homeland Security Advisory System. Please visit www.redcross.org/services/disaster/beprepared/hsas.html

And finally, for more information about the **Department of Homeland Security** and to stay current on national security alerts and warnings, please visit www.dhs.gov

# APPENDIX B

## Citizen Corps / CERT / CERV Ontario
### (Volunteer Programs for Americans & Canadians)

### What is Citizen Corps?

**Citizen Corps**, a component of USA Freedom Corps, was created to help coordinate volunteer activities that make the nation's communities safer, stronger, and better prepared to respond to any emergency situation.

Citizen Corps is managed at local levels by Citizen Corps Councils, which bring together existing crime prevention, disaster preparedness, and public health response networks with the volunteer community and other groups. Councils organize public education on disaster mitigation and preparedness, training, and volunteer programs for people of all ages and backgrounds.

### What programs are under Citizen Corps?

Citizen Corps programs (like the ones listed below) and many other projects and events are coordinated by local, state, and tribal Citizen Corps Councils.

- Volunteers in Police Service (VIPS) programs
- Community Emergency Response Teams (CERTs)
- Medical Reserve Corps units
- Neighborhood Watch groups

To learn more about **Citizen Corps** or to check if there is a local council in your community, please visit www.citizencorps.gov or www.fema.gov

### What is CERT or CERV?

In the United States and Canada, the **Community Emergency Response Team (CERT)** program helps train volunteers to assist first responders in emergency situations in their communities. CERT members give critical support to first responders in emergencies, provide immediate assistance to victims, organize spontaneous volunteers at a disaster site, and collect disaster intelligence to support first responder efforts.

For more information about CERT programs or to check if a CERT is in your community, visit http://training.fema.gov/emiweb/cert/index.asp ... or visit www.cert-la.com ... or call your local, State, Provincial, or Territorial Emergency Management Office to ask about volunteer opportunities.

The **Community Emergency Response Volunteers (CERV) Ontario** is co-ordinated through Emergency Measures Ontario. Visit the CERV Ontario site at www.mpss.jus.gov.on.ca/english/pub_security/emo/cerv_ont.html

# APPENDIX C

*The following **SAMPLE GUIDELINE / DATABASE** was developed and contributed by South Carolina's Charleston County Emergency Preparedness Department. A list of acronyms used in the following 6 pages is included on page 216.*

## Any County Emergency Preparedness
## Terrorism Emergency Operations Outline

### Counteractions Standard Operations Guide

## I. GENERAL

A. *Purpose*

The purpose of this SOG is to assist other officials and emergency service personnel with a working outline for developing a written set of guidelines for the conduct of antiterrorism and terrorism counteraction response operations. Contact your local emergency management agency for assistance.

This outline of antiterrorism is designed to deter and limit the success of terrorists acts against government / industry resources / personnel and facilities while Counteraction facilitates response to, and recovery from, an actual terrorist incident. The collection and dissemination of timely threat intelligence information, informative public awareness programs, and through the implementation of sound defensive mitigation measures usually insure the best countermeasures one may accomplish.

B. *Authority*

The listing of local city, county, state or federal ordinances as may be applicable for the intended jurisdictions being protected.

C. *References*

1. Presidential Decision Directive 39 [PDD-39], June 1995.

2. State Terrorism Incident Annexes.

3. Local Community Bomb Threat Incident Plan.

4. Emergency Response To Incidents Involving Chemical and Biological Warfare Agents.

5. Terrorism In The U.S. 1982-1992, FBI Report.

6. The Federal Response Plan [FRP], Terrorism Annex.

7. Local Airport, Seaport, Transportation, Dam, or Utilities Emergency Counter Terrorism Plan.

8. Emergency Response to Terrorism Job Aid, May 2000.

9. II CT Chemical / Biological Incident Handbook.

D. *Definitions*

A list as detailed or as brief as may be applicable to the depth of your SOG. Contact your local emergency management agency for assistance.

E. *Organizations*

Apply a basic organization chart for primary agencies that may be anticipated to support the various aspects of your SOG. Federal, state, city / county and local. See the local emergency management agency plan.

## II. SITUATION

A. *International Terrorism*

International terrorism continues as a threat and although it has been primarily rooted in other countries, a great number of Americans have been affected by these acts. According to the U.S. Department of State and as reported in the media large numbers of the actual terrorist attacks worldwide were targeted against Americans. The impact of international terrorism still has vivid images occurring almost worldwide.

B. *Domestic Terrorism*

Statistics provided by the FBI prior to the mid-1980's indicated only a few acts of terrorism inside the United States as influenced by international terrorism. Since the 1993 bombing of the World Trade Center in New York it was a clear reminder that the United States is not immune from acts of international terrorism based within our borders.

It is very clear that managing the consequences of terrorism in the United States from any source can be a most difficult and challenging task. Trying to recover from such senseless terrorist events has already begun to change the way Americans view the potential threats and mass effect from a single terrorist act perpetrated in local communities.

## C. *Conditions*

1. Actual events or threat of a terrorist act may cause implementation of precautionary measures from as high up as Presidential sources.

2. The FBI will likely implement a crisis management law enforcement response to any significant threat or actual act of terrorism and include threat assessment / consultation and NBC / WMD Technical Assistance.

3. Incidents that occur without advance threat or warning and that produce major consequences. FEMA will probably respond and implement within the FRP consequence management activities.

## D. *Planning Assumptions*

No single agency at any government level has the unilateral authority or all the knowledge and skills to act in a terrorist event, especially WMD / WME. The SOG will be activated upon such threat or an actual event.

Certain instances will require, as in NBC events, that perimeters be set and closed to authorized officials and first responders as well. The SOG may also have to request activation of specialty service resources and/or task forces. Your local emergency management plans may assist this step.

## III. CONCEPT OF OPERATIONS

## A. *Crisis Management*

1. Local Law Enforcement generally has lead responsibility for implementing SOG crisis management.

2. Each participating agency will maintain a current copy of the Terrorism Response Alert List of the SOG.

3. A systematic scene approach will often be implemented while self-protective measures as appropriate are taken towards controlling the situation. The IMS/ICS framework

and possible transition into the Unified Command System may be used as soon as possible.

4. Responders SITREP, staging, direction and command & control information all within often dynamic incident events as agencies / personnel and equipment arrive in force.

5. Communications size-up will be CONTINUOUS in such a dynamic incident and should address scene stability in the SOG (e.g. stable, deteriorated, continuing to deteriorate, unsafe).

6. To avoid infrastructure gridlock, establish from the initial SITREP the priorities needed for life safety and protection. Address immediate and sequential response structuring.

## B. *Consequence Management*

This level of management integrates all aspects of the response that will generally protect the public's health and safety, manage fears and suffering, and enhance evidence-gathering towards identifying and eventually apprehending the perpetrators. For assistance contact your local emergency management agency.

1. Pre-Incident Phase

   a. Protective actions such as organizational SOG's aimed at coordinating any threat in the local area via an identified part of a Command Group. Threatcon Alpha or Bravo.

   b. SOG's that establish actions and security awareness measures that prepare a counter deterrence to terrorist vulnerabilities.

2. Trans-Incident Phase

   a. This phase involves the threat emerging to an actual act or imminent action of terrorism. Threatcon Charlie.

   b. Everyone should stay focused on the end objective to "save lives" and coordinate cooperative agencies' efforts to solve most disagreements. ROC's, JOC's, JIC's, EOC's, IST's, IAP's, SOG.

3. Post-Incident Phase

   a. This phase may involve an incident that occurred without any advance warning and produces major consequences and appears to be an act of terrorism. Many concurrent efforts of crisis management will be initiated to establish a short-term Incident Action Plan.

b. Local officials will mitigate the situation to the best of their ability until further supported by the combined state and federal resources tasked.

4. Disengagement

a. If no act of terrorism occurs then the federal response will disengage as coordinated. Stand down will occur for all according to their SOP's / SOG's.

b. All agencies that responded will be requested to turn in a copy of their incident logs, journals, messages, or other non-sensitive records to the local coordinating agency. This information will be key to establishing an accurate post incident critique. Critiques are often delayed pending any legal requirements to keep certain information in confidence.

c. PISD (post incident stress debriefings) will be offered by the proper mental health agency for responders based on the nature of the event and it's circumstances.

## IV. RESPONSIBILITIES

A. *Crisis and Consequence Management*

1. The County Sheriff LNO will: develop local SOG's.

2. Local Response Agencies will: develop local SOG's.

B. *EOC - JOC Support Agencies*

Agencies not covered in this SOG are understood for mutual aid response to assist neighboring communities. See attached map.

*(Suggestion: Include a map with your plan indicating areas being covered by Support Agencies.)*

## V. LOCAL STATE FEDERAL INTERFACE

This SOG is supported by the Terrorism Incident Annex to the Federal Response Plan and your State Emergency Operations Plan to include any Regional Task Force operational concepts applicable.

Contact your local Emergency Management Agency for assistance in available plans.

## VI. PRIMARY POINT OF CONTACT

Inquiries or changes concerning this SOG Outline should be addressed to Charleston County EPD, Project Officer for Terrorism Incident Management, 4045 Bridge View Drive, North Charleston, S.C. 29405-7464 or 843-202-7400 and Fax 843-202-7408.

DISCLAIMER: Information provided is solely intended as a sample guideline / database and neither the County of Charleston nor any agency, officer or employee warrants the accuracy, reliability or timeliness of any information in the Terrorism Counteraction SOG database. While every effort is made to ensure a broad accuracy of this information, portions may be incorrect or not current for all circumstances and we shall not be liable for any losses caused by such reliance on this outline information. Any persons or entities who relies on information obtained from this database does so at his or her own risk.

# Acronyms

Used in preceding 6-page "**County Emergency Preparedness Terrorism Emergency Operations Outline**"

B-NICE - Biological, Nuclear Incendiary, Chemical or Explosive Device

CAT - Crisis Action Team

CBR - Chemical, Biological, Radiological

EOC - Emergency Operations Center

EPD - Emergency Preparedness Division

FEMA - Federal Emergency Management Agency

FRP - Federal Response Plan

IAP - Incident Action Plan

ICS - Incident Command System

IMS - Incident Management System

IST - Incident Support Team

JIC - Joint Incident Command

JOC - Joint Operations Center

LNO - LIAISON Officer

NBC - Nuclear, Biological, Chemical devices

PISD - Post Incident Stress Debriefings

ROC - Regional Operations Center

SITREP - Situation Report

SOP - Standard Operating Procedures

SOG - Standard Operating Guidelines

Threatcon - Terrorist Threat Condition

WME - Weapons of Mass Effect

WMD - Weapons of Mass Destruction

# END NOTES

[1] U. S. Department of Commerce, NOAA, Pacific Marine Environmental Laboratory, Tropical Atmosphere Ocean, "Frequently Asked Questions about El Niño and La Niña", (www.pmel.noaa.gov/tao/elnino/faq.html), 2004.

[2] NOAA News Online (Story 2317), "NOAA Announces the Return of El Niño", (www.noaanews.noaa.gov/stories2004/s2317.htm), September 10, 2004.

[3] NOAA "Answers to La Niña frequently asked questions", (www.elnino.noaa.gov/lanina_new_faq.html), May 3, 2001.

[4] Adam Entous, "Bush Signs Measure Boosting U.S. Bioterror Defenses" (Reuters online, June 12, 2002).

[5] Centers for Disease Control Public Health Emergency Preparedness & Response, Agents, Diseases, & Threats, Radiation Emergencies, Information for the Public, "Dirty Bombs", (www.bt.cdc.gov/radiation/dirtybombs.asp), 2003.

[6] Water Supply and Sanitation Collaborative Council, "WASH Facts and Figures", (www.wsscc.org), Geneva, Switzerland, 2002.

# RESOURCES

American Heart Association. *Guidelines 2000 for Cardiopulmonary Resuscitation and Emergency Cardiovascular Care, International Consensus on Science*, Volume 11, Number 3, Fall 2000.

American Medical Association. Journals of the AMA, JAMA Consensus Statement, *Tularemia as a Biological Weapon Medical and Public Health Management*, Chicago, IL, 2001.

American Red Cross. *Are You Ready for a Heat Wave?* Washington, D.C.: The American National Red Cross, 1998.

American Red Cross. *Are You Ready for a Thunderstorm?* Washington, D.C.: The American National Red Cross, 1998.

American Red Cross. *Coping With Disaster – Emotional Health Issues for Victims*, Washington, D.C.: The American National Red Cross, 1991.

American Red Cross. *Disaster Preparedness for Seniors by Seniors*, Washington, D.C.: The American National Red Cross and the Rochester-Monroe County Chapter of the American Red Cross, 1995.

American Red Cross. *Family Disaster Plan and Personal Survival Guide*, Washington, D.C.: The American National Red Cross, 1989.

American Red Cross. *First Aid Fast*, Washington, D.C.:The American National Red Cross, 1995.

American Red Cross and California Community Foundation. *Disaster Preparedness for Disabled & Elderly People*, Washington, D.C.: The American National Red Cross, 1985.

American Red Cross Los Angeles Chapter, *The Emergency Survival Handbook*, Los Angeles, CA: Los Angeles Chapter, American Red Cross, 1985.

American Red Cross Los Angeles Chapter – Third Edition, *Safety and Survival in an Earthquake,* Los Angeles, CA: L. A. Chapter, American Red Cross, 1986.

Centers for Disease Control and Prevention National Center for Environmental Health, Air Pollution & Respiratory Health, *Asthma*, Atlanta, GA, 03/04/2003.

Centers for Disease Control and Prevention National Center for Environmental Health, Air Pollution & Respiratory Health, *Carbon Monoxide*, Atlanta, GA, 06/06/2002.

Center for Disease Control and Prevention National Center for Infectious Diseases Division of Vector-Borne Infectious Diseases, *What Everyone Should Know About SARS (Severe Acute Respiratory Syndrome)*, Atlanta, GA, 08/27/2003.

Center for Disease Control and Prevention National Center for Infectious Diseases, Division of Vector-Borne Infectious Diseases, *West Nile Virus - Background*, Atlanta, GA, 11/13/2002.

Center for Disease Control and Prevention National Center for Infectious Diseases, Division of Vector-Borne Infectious Diseases, *West Nile Virus - Basics*, Atlanta, GA, 11/13/2002.

Center for Disease Control and Prevention National Center for Infectious Diseases, Division of Vector-Borne Infectious Diseases, *West Nile Virus - Questions and Answers*, Atlanta, GA, 08/29/2002.

Center for Disease Control and Prevention National Center for Infectious Diseases, Special Pathogens Branch. Disease Information, *Viral Hemorrhagic Fevers*, Atlanta, GA, 01/29/2002.

Centers for Disease Control and Prevention Public Health Emergency Preparedness and Response, *Children and Anthrax: A Fact Sheet for Parents*, Atlanta, GA, 11/07/2001.

Centers for Disease Control and Prevention Public Health Emergency Preparedness and Response, *Facts about Anthrax*, Atlanta, GA, 10/14/2001.

Centers for Disease Control and Prevention Public Health Emergency Preparedness and Response, *Facts about Botulism*, Atlanta, GA, 10/14/2001.

Centers for Disease Control and Prevention Public Health Emergency Preparedness and Response, *Facts about Chlorine*, Atlanta, GA, 03/18/2003.

Centers for Disease Control and Prevention Public Health Emergency Preparedness and Response, *Facts about Cyanide*, Atlanta, GA, 03/12/2003.

Centers for Disease Control and Prevention Public Health Emergency Preparedness and Response, *Facts about Pneumonic Plague*, Atlanta, GA, 10/14/2001.

Centers for Disease Control and Prevention Public Health Emergency Preparedness and Response, *Facts about Sarin*, Atlanta, GA, 03/07/2003.

Centers for Disease Control and Prevention Public Health Emergency Preparedness and Response, *Facts about Sulfur Mustard*, Atlanta, GA, 03/12/2003.

Centers for Disease Control and Prevention Public Health Emergency Preparedness and Response, *Facts about Tularemia*, Atlanta, GA, 02/06/2002.

Centers for Disease Control and Prevention Public Health Emergency Preparedness and Response, *Facts about VX*, Atlanta, GA, 03/12/2003.

Centers for Disease Control and Prevention Public Health Emergency Preparedness and Response, *Frequently Asked Questions (FAQ) About Plague*, Atlanta, GA, 10/3/2002.

Centers for Disease Control and Prevention Public Health Emergency Preparedness and Response, *Frequently Asked Questions (FAQs) About Ricin*, Atlanta, GA, 03/12/2003.

Centers for Disease Control and Prevention Public Health Emergency Preparedness and Response, *Other Frequently Asked Questions (FAQs) about Preparedness and Response and Water Safety*, Atlanta, GA, 10/25/2001.

Centers for Disease Control and Prevention Public Health Emergency Preparedness and Response, Radiation Emergencies, Background Information, *Acute Radiation Syndrome*, Atlanta, GA, 04/07/2003.

Centers for Disease Control and Prevention Public Health Emergency Preparedness and Response, Radiation Emergencies, Background Information, *Nuclear Terrorism & Health Effects*, Atlanta, GA, 04/07/2003.

Centers for Disease Control and Prevention Public Health Emergency Preparedness and Response, Radiation Emergencies, Emergency Instructions for Individuals & Families, *Dirty Bombs*, Atlanta, GA, 04/07/2003.

Centers for Disease Control and Prevention Public Health Emergency Preparedness and Response, *Smallpox Fact Sheet*, Atlanta, GA, 12/09/2002.

De Blij, H. J., *Nature on the Rampage*, Smithsonian Institute, Washington D.C.: Smithsonian Books, 1994.

Department of Health & Human Services, National Institutes of Health, U.S. National Library of Medicine, Specialized Information Services. Toxicology & Environmental Health, Special Topics, *Biological Warfare*, Bethesda, MD, 02/18/2003.

Editors of BACKPACKER® Magazine, *All Weather All Season Trip Planner*, Emmaus, PA:Rodale Press, Inc., 1997.

Emergency Preparedness Canada, British Columbia Provincial Emergency Program, Canadian Mortgage and Housing Corporation, Health Canada, Geological Survey of Canada, Insurance Bureau of Canada, *Earthquakes in Canada?*, Her Majesty the Queen in Right of Canada, Department of Natural Resources Canada, 1996.

Emergency Preparedness Canada, Canadian Geographic, Environment Canada, Geological Survey of Canada, Insurance Bureau of Canada, Natural Resources Canada, Statistics Canada, The Weather network, *Natural Hazards - a National Atlas of Canada*, Canadian Services Canada, 1997.

Federal Emergency Management Agency, *Are You Ready? A Guide to Citizen Preparedness,* Washington, D.C., December 2002.

Federal Emergency Management Agency, *Are You Ready? Your Guide to Disaster Preparedness,* Washington, D.C., 1993.

Federal Emergency Management Agency, *Extreme Heat Fact Sheet & Backgrounder*, Washington, D.C., 1998.

Federal Emergency Management Agency, *Good Ideas Book – How People and Communities Are Preparing For Disaster*, Washington, D.C., 1993.

Federal Emergency Management Agency, *Hazardous Materials Fact Sheet & Backgrounder*, Washington, D.C., 1998.

Federal Emergency Management Agency, *The Humane Society of the United States Offers Disaster Planning Tips for Pets, Livestock and Wildlife*, Washington, D.C., 1997.

Federal Emergency Management Agency, *Landslides and Mudflows Fact Sheet & Backgrounder*, Washington, D.C., 1998.

Federal Emergency Management Agency, *National Flood Insurance Coverage Available to Homeowners*, Washington, D.C., 1996.

Federal Emergency Management Agency, *Nuclear Power Plant Emergency Fact Sheet & Backgrounder*, Washington, D.C., 1997.

Federal Emergency Management Agency, *Returning Home After the Disaster – An Information Pamphlet for FEMA Disaster Workers*, Washington, D.C., 1987.

Federal Emergency Management Agency, *Terrorism Fact Sheet & Backgrounder*, Washington, D.C., 11-Feb-2003.

Federal Emergency Management Agency, *Volcanoes Fact Sheet & Backgrounder*, Washington, D.C., 1998.

Federal Emergency Management Agency and the American Red Cross, *Disaster Preparedness Coloring Book*, Washington D.C., 1993.

Federal Emergency Management Agency and the American Red Cross, *Emergency Preparedness Checklist*, Washington D.C., 1993.

Federal Emergency Management Agency and the American Red Cross, *Food and Water in an Emergency*, Washington D.C., 1994.

Federal Emergency Management Agency and the American Red Cross, *Helping Children Cope with Disaster*, Washington D.C., 1993.

Federal Emergency Management Agency and the American Red Cross, *Preparing for Emergencies - A Checklist for People with Mobility Problems*, Washington D.C., 1995.

Federal Emergency Management Agency and the American Red Cross, *Your Family Disaster Supplies Kit,* Washington D.C., 1992.

Federal Emergency Management Agency, the American Red Cross, The Home Depot, National Association of Home Builders, Georgia Emergency Management Agency, *Against the Wind - Protecting Your Home from Hurricane Wind Damage,* Washington D.C., 1993.

Federal Emergency Management Agency and the United States Fire Administration, *Wildfire: Are You Prepared?* Washington D.C., 1993.

Federal Emergency Management Agency and the Wind Engineering Research Center at Texas Tech University, *Taking Shelter From the Storm Building a Safe Room Inside Your House, Second Edition,* Washington D.C., August 1999.

Grant, Ashley H., "Mosquito experts discuss West Nile Fear", Associated Press, 03/04/2003.

Harvard Center for Risk Analysis, "Using Decision Science to Empower Informed Choices About Risks to Health, Safety, and the Environment", Boston, MA, 2003.

Health Canada, Diseases & Conditions, *West Nile Virus*, Ottawa, Ontario, Canada, 2003-01-15.

Health Canada, Emergency Preparedness and Response, *Anthrax*, Ottawa, Ontario, Canada, 2002-04-18.

Health Canada, Emergency Preparedness and Response, *Botulism*, Ottawa, Ontario, Canada, 2002-11-18.

Health Canada, Emergency Preparedness and Response, *Smallpox*, Ottawa, Ontario, Canada, 2003-02-07.

Health Canada, Emergency Preparedness and Response, *The Plague*, Ottawa, Ontario, Canada, 2002-04-18.

Health Canada, Emergency Preparedness and Response, *Tularemia (Rabbit fever)*, Ottawa, Ontario, Canada, 2002-04-18.

Health Canada, Emergency Preparedness and Response, *Viral Haemorrhagic Fevers*, Ottawa, Ontario, Canada, 2002-04-18.

Health Canada, Warnings / Advisories, *Questions and Answers - Severe Acute Respiratory Syndrome (SARS)*, Ottawa, Ontario, Canada, 2003-04-15.

Information on avalanches obtained from the Internet online information page, "Avalanche Awareness", (http://nsidc.org/snow/avalanche/) maintained by the National Snow and Ice Data Center, University of Colorado Cooperative Institute for Research in Environmental Sciences, Boulder, November 2000. *(Revised URL 2003)*

Kozaryn, Linda D. "Defending Against Invisible Killers - Biological Agents", American Forces Press Service, March 1999, Department of Defense.

Mayell, Mark and the Editors of *Natural Health* Magazine, *The Natural Health First-Aid Guide: the definitive handbook of natural remedies for treating minor emergencies*, New York: Pocket Books, 1994.

Microsoft Corporation. "Meteorology," Microsoft® Encarta® Online Encyclopedia 2001 http://encarta.msn.com © 1997-2001.

National Institutes of Health, National Institute on Deafness and Other Communication Disorders, Health Information, Hearing, Ear Infections, and Deafness, *Noise-Induced Hearing Loss*, Bethesda, MD, September 2002.

National Oceanic and Atmospheric Administration, National Weather Service, West Coast & Alaska Tsunami Warning Center. *Physics of Tsunamis*, 02/20/2003.

National Oceanic and Atmospheric Administration, National Weather Service, West Coast & Alaska Tsunami Warning Center. *Tsunami Safety Rules*, 02/20/2003.

Nova Scotia Museum, An Illustrated Guide to Common Nova Scotian Poisonous Plants, Poisonous Leafy Plants, *Castor Beans*, Halifax, Nova Scotia, Canada, 00-03-08.

OCIPEP, Disaster Mitigation, *Frequently Asked Questions*, Ottawa, Ontario, Canada, 12/13/2002.

OCIPEP, Disaster Mitigation, *Towards a National Disaster Mitigation Strategy*, Ottawa, Ontario, Canada, 12/13/2002.

OCIPEP, Emergency Preparedness, *About Emergency Preparedness*, Ottawa, Ontario, Canada, 12/10/2002.

OCIPEP, Information Products, Facts Sheets, *Canada's Emergency Management System*, Ottawa, Ontario, Canada, 01/30/2003.

Pima County Health Department, Communicable Diseases / Bioterrorism, *Pima County Prepares for Smallpox - Frequently Asked Questions,* Tucson, Arizona, 2003.

Reader's Digest, *Natural Disasters (The Earth, Its Wonders, Its Secrets)*, London: The Reader's Digest Association, Limited, 1996.

State of California Department of Conservation and National Landslide Information Center, U.S. Geological Survey, "Features That May Indicate Catastrophic Landslide Movement", Denver, November 1998.

Survivor Industries, Inc., "The Wallace Guidebook for Emergency Care and Survival", Newbury Park, CA: H. Wallace, 1989.

Swift Aid USA, *SwiftAid to First Aid "What To Do In An Emergency"*, Toledo, OH, 1990.

United Nations Environment Programme, "Global Warming Report Details Impacts On People and Nature", Bonn/Geneva/Nairobi, 19Feb2001.

United States Postal Service®, United States Postal Inspection Service, "Notice 71 - Bombs By Mail", Chicago, IL, February 1998.

University of Colorado at Boulder, *Natural Hazards Observer*, "Congress Creates Department of Homeland Security", Volume XXVII Number 3, January 2003.

U.S. Army's Office of the Surgeon General's Medical NBC Online. Medical References Online, "Chemical Agent Terrorism" by Frederick R. Sidell, M.D. of the U.S. Army Medical Research Institute of Chemical Defense, (no date shown on www.nbc-med.org).

U.S.D.A. Forest Service National Avalanche Center, "Avalanche Basics": www.avalanche.org/~nac/ , November 2000.

U.S. Department of the Interior, U.S. Geological Survey, Cascades Volcano Observatory and Washington State Military Department, Emergency Management Division, *What To Do If A Volcano Erupts Volcanic Ashfall - How to be Prepared for an Ashfall*, Vancouver, WA, November 1999.

U.S. Department of the Interior, U.S. Geological Survey Earthquake Hazards Program, For Kids Only, *Cool Earthquake Facts*, Menlo Park, CA, 05-Mar-2003.

U.S. Department of the Interior, U.S. Geological Survey Volcano Hazards Program, *Effects of Lahars*, Menlo Park, CA, 10/15/1998.

U.S. Department of the Interior, U.S. Geological Survey Volcano Hazards Program, *Eruption Warning and Real-Time Notifications*, Menlo Park, CA, 01/30/2001.

U.S. Department of the Interior, U.S. Geological Survey Volcano Hazards Program -- *Reducing volcanic risk, Photo Glossary of volcano terms*, Menlo Park, CA, 01/30/2001.

U.S. Department of the Interior, U.S. Geological Survey Volcano Hazards Program, *Pilot Project Mount Rainier Volcano Lahar Warning System*, Menlo Park, CA, 09/04/2000.

U.S. Department of the Interior, U.S. Geological Survey Volcano Hazards Program, *Tephra: Volcanic Rock and Glass Fragments*, Menlo Park, CA, 12/23/1999.

U.S. Environmental Protection Agency, Indoor Air Quality, *Molds*, Washington, D.C., 03/03/2003.

U.S. Environmental Protection Agency, EPA Newsroom, "EPA Administrator Whitman Urges Home Testing for Radon, Commemorates National Radon Action Month", Washington, D.C., 01/14/2003.

Wellesley College, Web of Species "Jewelweed, Spotted Touch-Me-Not", Boston, MA, www.wellesley.edu/Activities/homepage/web/Species/ptouchmenot.html , 01/31/2001.

Worldwatch Institute, Abramovitz, Janet N., "Human Actions Worsen Natural Disasters", Press Release for Worldwatch Paper 159, Washington, DC, www.worldwatch.org, 10/18/2001.

Worldwatch Institute, Abramovitz, Janet N., "Natural disasters – At the hand of God or man?", Environmental News Network (ENN) Features, 23 June 1999, Copyright 1999.

# Additional Resources
# & Web Sites

## American Red Cross Disaster Services:

*After Disaster Strikes – Recovering financially*
The American Red Cross, the Federal Emergency Management Agency, and the National Endowment for Financial Education published the original brochure. www.redcross.org/pubs/

*Talking About Disaster: Guide for Standard Messages*
The Guide is a set of standard disaster safety messages on many hazards as well as general disaster safety information and viewable through web pages or using downloadable PDF files. Members of the National Disaster Education Coalition include the American Red Cross, FEMA, NOAA/National Weather Service, National Fire Protection Association, U.S. Geological Survey, Institute for Business and Home Safety, International Association of Emergency Managers, and the U.S. Department of Agriculture Cooperative State Research, Education, and Extension Service. www.redcross.org/disaster/safety/guide.html

## Federal Emergency Management Agency (FEMA):

**FIMA** (Federal Insurance & Mitigation Administration) www.fema.gov/fima

To order FEMA materials call local or state EM office. Or call FEMA: 1-800-480-2520 M-F 8a-5p EST, Fax 301-497-6378. Or visit www.fema.gov (most materials online)

## Health Canada's Emergency Preparedness:

**Health Canada**'s role in an emergency, whether it's a natural disaster or human caused, is to protect the health of Canadians. www.hc-sc.gc.ca/english/epr

**Centre for Emergency Preparedness and Response (CEPR)**
Serves as Canada's single coordinating point for public health emergencies. The 24-hour Centre works closely with experts in areas such as infectious disease, food and blood safety, nuclear, radiological, biological and chemical threats, and many other preparedness and response issues through **4 specialized offices**:

> **Office of Emergency Preparedness, Planning and Training**
> **Office of Emergency Services**
> **Office of Laboratory Security**
> **Office of Public Health Security**

**CEPR** www.hc-sc.gc.ca/pphb-dgspsp/cepr-cmiu/cepr.html

Through the above **CEPR** offices, Health Canada is also responsible for the National Emergency Stockpile System (NESS), the Emergency Social Services (ESS), the Federal Nuclear Emergency Plan (FNEP), providing emergency health care for First Nations and Inuit communities, the health of travelers entering Canada, and many other health-related functions. www.hc-sc.gc.ca

PUBLIC SAFETY AND EMERGENCY PREPAREDNESS CANADA (FORMERLY OCIPEP) IN TRANSITION: www.ocipep.gc.ca and www.psepc-sppcc.gc.ca

**Alerts & Advisories** www.ocipep.gc.ca/opsprods/index_e.asp

**Critical Infrastructure** www.ocipep.gc.ca/critical/index_e.asp

**Emergency Preparedness** www.ocipep.gc.ca/ep/index_e.asp

**Information Products** www.ocipep.gc.ca/info_pro/index_e.asp


MISCELLANEOUS SITES *(* = COOL STUFF FOR EDUCATORS, KIDS & SCHOOLS)*

* **American Academy of Pediatrics** - Children, Terrorism & Disasters www.aap.org/terrorism

**American Avalanche Association** www.americanavalancheassociation.org

**American Stroke Association** www.strokeassociation.org

**Canada Mortgage and Housing Corporation** www.cmhc.ca

**Canadian Avalanche Association** www.avalanche.ca

**Canadian Centre for Emergency Preparedness** www.ccep.ca

* **Canadian Network of Toxicology Centres, University of Guelph** - Project Earth Risk Identification Life www.uoguelph.ca/cntc/educat/peril.htm

* **Canadian Red Cross** www.redcross.ca

* **Centers for Disease Control and Prevention National Center for Environmental Health** www.cdc.gov/nceh

**Centers for Disease Control and Prevention Public Health Emergency Preparedness and Response** www.bt.cdc.gov

**Disaster Recovery Information Exchange (DRIE) Canada** www.drie.org

* **Environment Canada** www.ec.gc.ca *(Check out their Topics)*

**Environmental Protection Agency (EPA)** www.epa.gov

* **EPA's Explorers' Club for Kids** www.epa.gov/kids

**EPA's Chemical Emergency Preparedness and Prevention Office (CEPPO)** http://yosemite.epa.gov/oswer/ceppoweb.nsf/content/index.html

* **FEMA for Kids** www.fema.gov/kids

* **Harvard Center for Risk Analysis** www.hcra.harvard.edu

**Heart and Stroke Foundation of Canada** www.heartandstroke.ca

**Humane Society Disaster Services Program** www.hsus.org/disaster

**Institute for Business and Home Safety** www.ibhs.org

* **Institute for Catastrophic Loss Reduction** (Toronto, Ontario) www.iclr.org

**Insurance Bureau of Canada** www.ibc.ca

* **Lyme Disease Foundation** www.lyme.org

**Munich Re Group**'s Press Releases www.munichre.com

**National Fire Protection Association** www.nfpa.org
 * Fire Wise (Info on wildfires) www.firewise.org
 * Risk Watch www.nfpa.org/riskwatch
 * Sparky the Fire Dog www.nfpa.org/sparky

**National Interagency Fire Center** (Boise, Idaho) www.nifc.gov

* **Natural Resources Canada** www.nrcan.gc.ca

**NOAA (National Oceanic and Atmospheric Administration)** www.noaa.gov
 Atlantic Oceanographic & Meteorological Laboratory www.aoml.noaa.gov
 National Climatic Data Center www.ncdc.noaa.gov/ol/ncdc.html
 National Weather Service (NWS) www.nws.noaa.gov
 * NWS Lightning Safety www.lightningsafety.noaa.gov
 * NWS Public Affairs Links for Kids www.nws.noaa.gov/pa/forkids.html
 * Storm Prediction Center Online Tornado FAQ www.spc.noaa.gov/faq/tornado

* **OCIPEP Teacher Corner** www.ocipep.gc.ca/info_pro/teachers/index_e.asp

**Pacific Tsunami Warning Center** (Hawaii) www.nws.noaa.gov/pr/ptwc

* **PBS Online's SAVAGE EARTH** www.pbs.org/wnet/savageearth

**Royal Canadian Mounted Police** www.rcmp-grc.gc.ca

**Statistics Canada** www.statcan.ca

* **United Nations Environment Programme** www.unep.org

**U.S. Department of Defense DefenseLINK** www.defenselink.mil

* **U.S. Department of Education Office of Safe and Drug-Free Schools**
 Emergency Planning for Schools www.ed.gov/emergencyplan

**U.S. Department of Health & Human Services Disasters & Emergencies**
 www.hhs.gov/disasters

**U.S. Department of Homeland Security** www.dhs.gov

**U.S. Fire Administration** www.usfa.fema.gov

**U.S.D.A. Forest Service National Avalanche Ctr** www.fsavalanche.org

* **U.S.G.S. Earthquake Hazards Program** www.earthquake.usgs.gov

* **U.S.G.S. Volcano Hazards Program** http://volcanoes.usgs.gov

* **U.S. Nuclear Regulatory Commission** www.nrc.gov

* **Volcano World** http://volcano.und.nodak.edu/vw.html

**West Coast / Alaska Tsunami Warning Center** http://wcatwc.gov

**World Nuclear Association** www.world-nuclear.org

**World Meteorological Organization** (United Nations Agency) www.wmo.ch

**World Water Council** www.worldwatercouncil.org

* **World Wide Fund for Nature** www.wwf.org

**Worldwatch Institute** www.worldwatch.org

# Index

## A

activated charcoal, first aid uses for, 21, 142
air pollution, ozone alerts, 54
air quality, mitigation tips, 34
American Red Cross. *See also* Canadian Red Cross
  about, 188
  assistance following disasters, 124
  first aid services and programs, 134
amputation, emergency measures, 149
anthrax (biological agent), 83-84
  how spread, 83
  signs and symptoms of exposure, 84
  treatment, 84
asthma attack, first aid treatment, 150
avalanches. *See also* landslides
  basics, 41, 42-43
  facts and figures, 4
  safety information, 43-45
  types of, 41-42
  typical victims, 42

## B

baking soda, first aid uses for, 21
  burns, 155
  paste for bites or stings, 142, 145
  paste for rash, 177
biological agents, 80, 82-91. *See also* terrorism
  basic groups of, 82
  how used in an attack, 83
  safety information, 80, 82-91
    after an attack, 91
    before an attack, 89-90
    during an attack, 90-91
  types of, 83-89
    anthrax, 83-84
    botulism, 84
    plague, 85
    ricin, 85-86
    smallpox, 86-88
    tularemia, 88
    viral hemorrhagic fevers (VHFs), 88-89
  where to get more information, 91
bites
  animals or humans, first aid treatment, 141
  snakes, first aid treatment, 145-146
  ticks, first aid treatment, 146-147

emergency preparedness in Canada, 197-198
partnerships, 198-199
   Provincial & Territorial EMOs, alphabetic listing of, 198-199
regional offices, 196


**R**

radiation. *See also* radiological threat or event
  detection, 74
  reducing exposure to, 76
  sickness, first aid treatment and information, 105
  using potassium iodide (KI) pills, 74-75, 78, 105
radiological threat or event
  dispersion device (RDD or dirty bomb), 74, 101-102
  safety information, 101-105
    after an event or explosion, 104-105
    before a threat or event, 103-104
    during an event or explosion, 104
  where to get more information, 105
Red Crescent Societies, 200
Red Cross
  American, 188
  Canadian, 195
  first aid services and programs, 134-135
  International, 200
rescue breathing. *See* mouth-to-mouth resuscitation
resources, 217-225
ricin (biological agent), 85-86
  biotoxin (also considered a chemical agent), 85, 92
  how spread, 85-86
  signs and symptoms of exposure, 86
  treatment, 86


**S**

safe room, for protection against
  chemical attack, 99
  hazardous materials, 66
  wind, 32
sanitation
  disinfectants, 130
  disposal of human waste in disaster situation, 130
  supplies to help with, 23
sarin (chemical agent), 92, 95-96
  how spread, 95
  signs and symptoms of exposure, 95-96
  treatment, 96
SARS (Severe Acute Respiratory Syndrome)
  symptoms and reducing the spread of, 181
  where to get more information, 181
scorpion stings, first aid treatment, 146-147

# NOTES